Introduction to Real Analysis

Introduction to Real Analysis

Editor

Lavra Filipek

Introduction to Real Analysis

Edited by **Lavra Filipek**

Printed in 2017

ISBN: 978-1-68117-189-0

Library of Congress Control Number: 2015949231

© 2016 by
SCITUS Academics LLC,
616, Corporate Way, Suite 2, 4766,
Valley Cottage, NY 10989

www.scitusacademics.com

Preface

Real analysis is a branch of mathematical analysis dealing with the real numbers and real-valued functions of a real variable. In particular, it deals with the analytic properties of real functions and sequences, including convergence and limits of sequences of real numbers, the calculus of the real numbers, and continuity, smoothness and related properties of real-valued functions.

The real numbers have several important lattice-theoretic properties that are absent in the complex numbers. Most importantly, the real numbers form an ordered field, in which addition and multiplication preserve positivity. Moreover, the ordering of the real numbers is total, and the real numbers have the least upper bound property. These order-theoretic properties lead to a number of important results in real analysis, such as the monotone convergence theorem, the intermediate value theorem and the mean value theorem.

A sequence is usually defined as a function whose domain is a countable totally ordered set, although in many disciplines the domain is restricted, such as to the natural numbers. In real analysis a sequence is a function from a subset of the natural numbers to the real numbers.

This textbook is a reference tool for junior or senior mathematics majors and science students with a serious interest in mathematics.

Contents

Two New Iterated Maps for Numerical Nth Root Evaluation

Charles Corrêa Dias[1],
Fernanda Jaiara Dellajustina[2],
and Luciano Camargo Martins[2]
[1]Department of Electrical Engineering,
Universidade do Estado de Santa Catarina
(UDESC), Joinville, Brazil
[2]Department of Physics, Universidade do
Estado de Santa Catarina (UDESC), Joinville,
Brazil

ABSTRACT

In this paper we propose two original iterated maps to numerically approximate the nth root of a real number. Comparisons between the new maps and the famous Newton-Raphson method are carried out, including fixed point determination, stability analysis and measure of the mean convergence time, which is confirmed by our analytical convergence time model. Stability of solutions is confirmed by measuring the Lyapunov exponent over the parameter space of each map. A generalization of the second map is proposed, giving rise to a family of new maps to address the same problem. This work is developed within the language of discrete dynamical systems.

INTRODUCTION

Recent applications of iterated maps in numerical analysis have been found in literature, using and extending the techniques of dynamical systems to the study of numerical algorithms and number theory [1] -[3] . Application in technology and hardware devices are also frequent nowadays [4] - [7].

We propose and study in this work two new methods for numerical root approximations, both of which based on iterated maps. In the following sections we present a detailed study of each map, their fixed points and stability, the occurrence of bifurcations and chaotic behavior.

Some common tools of nonlinear dynamics [8] [9] are used to the study of the orbits, i.e., the numerical time series obtained for each map are investigated. The numerical approximations x_i, i=0, 1, 2, L to solve the general equation $x^n = k$, where $n \in N$ and $k \in +R$ are obtained, but the validity of the results can be extended to $n \in +R$.

In Section 2 we present a new map proposed by one of us (C. C. Dias), named as First Dias Map (FDM), showing the existence of a fixed point for roots in the range $n \in [1, 4]$, and that this fixed point corresponds exactly to $\sqrt[n]{k}$.

In Section 3, we generalize the FDM by adding a new parameter for studying the stability of its fixed point by defining a new class of maps called Weighted Average Map (WAM). For this class of maps, we investigate the dependence of the fixed point corresponding to the nth root of k over the parameter space (n,p).

Finally, in Section 4, we measure and compare the Mean Convergence Time (MCT) for all the studied maps, for n=2 and n=3, varying k over an uniform grid of initial conditions and computing the average amount of iterations to converge to the root $\sqrt[n]{k}$ within the standard numerical double precision. An analytical model is proposed and used to confirm the numerical results of MCT with the analytical convergence time (ACT) for the WAM.

THE FIRST DIAS MAP (FDM)

The map which we will study now was created by Charles C. Dias to extract real roots of numbers numbers, by solving the equation $x^n = k$. The proposed map is one-dimensional and is defined as

$$x_{t+1} = \frac{1}{2}\left(x_t + \frac{k}{x_t^{n-1}}\right).$$

$$(1.1)$$

Comparing this with the Newton-Raphson Method (NRM) equation and Babylonian Method (BABM) [10] noticed that this statement is a mixture of both, and the FDM is an arithmetic average between the linear and nonlinear terms in Equation (1.1), and can be used to approximate the nth root of k>0 for $n \in (0,4)$, as we shall see. Outside this interval of n, this map presents chaotic dynamics through after entering a bifurcation cascade, whose roots having no longer relationship to the nth root of k.

The base function that appears in the iterated map defined by Equation (1.1) can be derived dividing $x^n = k$ by x^{n-1}, for $x \neq 0$, leading to

$$x = \frac{k}{x^{n-1}},$$

(1.2)

and adding x at both sides, and dividing it by 2, we recover functional form of Equation (1.1).

Geometrical Construction

To construct geometrically the FDM time series, the first step is to find the auxiliary equations of the lines $\overline{A_i B}$ (see Figure 1), writing their slopes $m_i = \Delta y_i / \Delta x_i$. From this figure, $\Delta y_i = f(x_i) + k$ and $\Delta x_i = x_i$, and the slopes

$$m_i = \frac{f(x_i) + k}{x_i}, \quad i = 0,1,2,3,\cdots$$

(1.3)

that for $f(x) = x^n - k$ are $m_i = x_i^{n-1}$.

Knowing that their linear coefficients are all -k, then all the auxiliary lines pass through the point $B(0,-k)$, we obtain the working lines

$y(x) = x_i^{n-1} x - k$ and the auxiliary points S_i, the intersections points of the working lines with the x-axis, are

$$S_i = \frac{k}{x_i^{n-1}}, \quad i = 0, 1, 2, 3, \cdots,$$

(1.4)

and taking the arithmetic mean between the auxiliary points S_i and the x_i points we recover the original FDM equation (Equation (1.1)).

Figure 2(a) shows the cobweb for the FDM time series for n=2 and k=2, and Table 1 (top) shows time series used in this figure. In this example, the convergence to the root is achieved after only five steps, considering the standard double precision, and is exactly the same time series of NRM for these parameters. Figure 2(b) shows a numerical development of the FDM series for the parameters n=3 and k=2, based on the time series shown in Table 1 (bottom), where the convergence to the root occurs after 27 steps. Some intermediary time steps are omitted in this table.

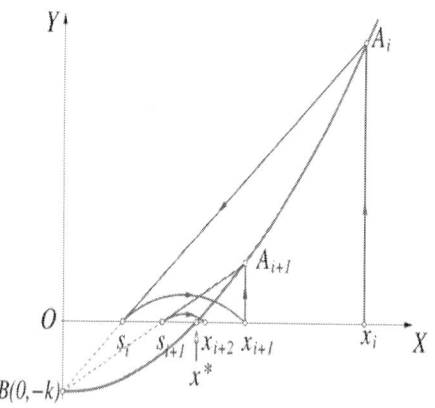

Figure 1: The FDM schematic geometrical path construction.

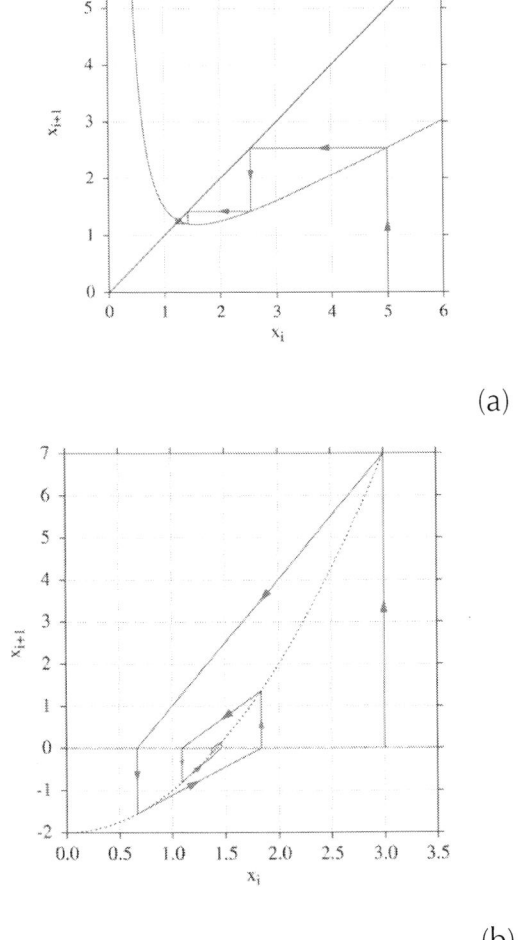

(a)

(b)

Figure 2: The FDM cobwebs for k=2: (a) n=2; (b) n=3.

Fixed Point and Stability Analysis

Solving $f(x^*) = x^*$ we find the FDM fixed point to be $x^* = \sqrt[n]{k}$. Applying the stability criterion [11], to the map function $f(x) = (x + k/x^{n-1})/2$, whose derivative is $f(x) = (1 - k(n-1)/x^n)/2$ we obtain $|1 - n/2| < 1$, and solving the last equation we have the range of parameter $n \in [0,4]$ where the fixed point of the map is stable.

Numerical Results

The FDM time series have different dynamics depending on the parameters (k,n), presenting a fixed point, periodicity or chaos, as occurs to the logistic map [12] .

To measure the rate of divergent orbits, i.e., the sensitive dependence on initial conditions, we can use is characteristic Lyapunov exponent $\lambda(k,n)$. From Figure 3(a) we see that, at n=4, FDM enters a bifurcation cascade, therefore its fixed point is no longer stable. In the white to gray regions the exponent is negative indicating that for this region of parameters the FDM not is chaotic, and at the black stripes the Lyapunov exponent goes to zero signing the period bifurcations. The yellow to red regions indicate the a positive Lyapunov exponents, the signature of chaos.

The FDM bifurcation diagram, discarded a transient of 10^3 iterations and plotted the next 500 values of x, is depicted in Figure 3(b). The values of n studied are uniformly distributed in a grid of 600 points in the interval [1, 10]. Also plotted over the bifurcation diagram is the exact root $\sqrt[n]{k}$, the fixed point of the map, plotted in black.

We also study numerically the FDM return diagrams for different values of the parameter n, for k=2 and $x_0 = 5$, as seen in Figure 3(c) and Figure 3(d), respectively.

THE WEIGHTED AVERAGE MAP (WAM)

Instead of adding x, if a more general term px is added in Equation (1.2), we get a new map (WAM) that depends on parameters n, k and p. The new parameter p is a positive real number and corresponds to the weight of the linear term of the map. This term is directly linked to

Two New Iterated Maps for Numerical Nth Root Evaluation

Table 1: FDM time series for k=2 for n=2 (top) and n=3 (bottom)

$n=2$

i	x_i	s_i	i	x_i	s_i
0	3.00000000	0.6666667	4	1.41421378	1.41421334
1	1.83333333	1.09090909	5	1.41421356	1.41421356
2	1.46212121	1.36787565	6	1.41421356	1.41421356
3	1.41499842	1.41342913	7	1.41421356	1.41421356

$n=3$

i	x_i	s_i	i	x_i	s_i
0	3.00000000	0.22222222	20	1.25992076	1.25992163
1	1.61111111	0.77051130	25	1.25992106	1.25992103
2	1.19081120	1.41040608	27	1.25992105	1.25992105
3	1.30060864	1.18232460	28	1.25992105	1.25992105

the parameter n, since for each value of n there is a minimum value of p for the fixed point to be stable, as we shall see. Adding px to both sides of Equation (1.2) we gain

$$px + x = px + \frac{k}{x^{n-1}}$$

(1.5)

and after collecting x and dividing by (p+1) it leads to the new map (WAM),

$$x = \frac{1}{p+1}\left(px + \frac{k}{x^{n-1}} \right),$$

(1.6)

and solving its fixed point equation $x_{i+1} = x_i = x^*$ we obtain the expected value $x^* = \sqrt[n]{k}$.

Fixed Point and Stability Analysis

Applying the stability criterion [11] , i.e., $\left| f'(x^*) \right| < 1$, to the map function $f(x) = (1/p+1)(px+k/x^{n-1})$ whose derivative is $f'(x) = (p-k(n-1)/x^n)/(p+1)$ and solving this inequality we obtain $p > n/2 - 1$ to guarantee fixed point stability, and Figure 4 shows the line corresponding to this condition, below which the fixed point is unstable. As n is increased the value of p should also be increased to avoid the unstable region, where the time series do not converges to the fixed point.

WAM Subclasses and Hierarchy

A special subclass of WAM is FDM, when p=1, so that the fixed point on the map according to Figure 4, is stable in the range [1,4] for n, thus in accordance with the stability analysis the fixed point $\sqrt[n]{k}$ loses stability diagram showing the stable fixed point for $n \in (0,4)$; return diagrams for $x_0=5$ and k=2: (c) n=2; (d) n=3.\

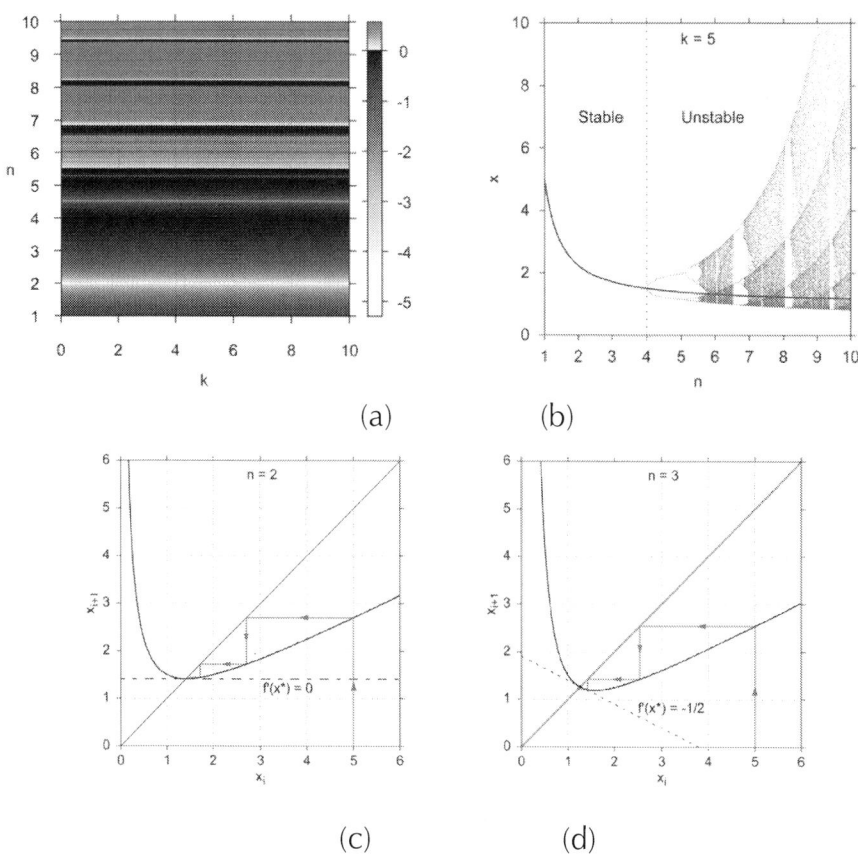

(a)   (b)

(c)   (d)

Figure 3: (a) FDM Lyapunov exponent over the parameter space (k,n); (b) the bifurcation

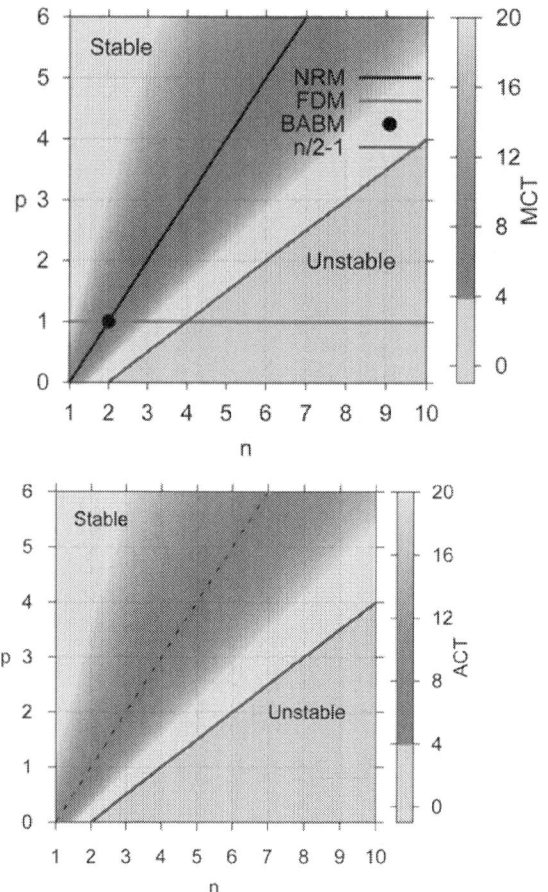

Figure 4: For k=2, (a) the numerical results of WAM MCT (n, p) and (b) the analytical model for WAM ACT (n, p).

At n=4. Other very important subclass is NRM, for p=(n-1), so that the weight p is chosen within the stable region.

For NRM, the derivative of the mapping function at the fixed point is null, satisfying the stability criterion and resulting in the most efficient rate of convergence of the time series near the fixed point. When the starting point x_0 is chosen far away from the fixed point, we observe numerically that the initial rate of convergence is greater for FDM, in general. Finally, there is a third subclass of WAM, the Babylonian square root method (BABM), for p=1 and n=2, the oldest and perhaps the most efficient known method to solve the equation $x^2=k$. At this parameters, WAM reduces itself to BABM, therefore the same occurs with FDM and NRM. The hierarchy of WAM subclasses are shown in Figure 5.

From the definition of the Lyapunov characteristic exponent for a uni-dimensional map we conclude that the derivative of the mapping function at the fixed point $f'(x^*)$ defines the rate of convergence of its time series, discarded the transient. For WAM, it is easy to show that $f'(x^*) = (p+1-n)/(p+1)$, so that this derivative assumes the value $1-n/2$ when p=1 (FDM) and is zero only if p=n-1 (NRM or BABM, if n=2). In Figure 3(c) we observe that $f'(x^*) = 0$, for n=2 FDM reduces to NRM, and in Figure 3(d), we observe that $f'(x^*) = -1/2$.

MEAN CONVERGENCE TIME (MCT)

This section reports the numerical results for the mean convergence time (MCT) for NRM, FDM and WAM, based on the average number of iterates to converge within different precisions ε, from single (10^{-8}) to double (10^{-16}). For this, we varied k on a uniform grid with 10^3 points in the interval $[10^{-1}, 10^2]$, varying the initial condition x_0 on a second uniform grid with 10^3 points, whose limits are given by a maximum relative difference of 25% around the exact value of the root of k at each point. Using this schema, the MCT is computed for cubic roots (n=3), and for WAM we set p=3. Figures 6(a)-(c) show the numerical results.

In Figure 6(a) we see that the NRM MCT is close to 4, which means that after 4 iterations, on average, there has been convergence to the root. From this figure, we conclude that FDM is around 10 times slower than NRM, and WAM is around has twice the speed of FDM. In this test, the most efficient is NRM, with the lowest MCT.

Both NRM and FDM belong to the same WAM family, as discussed in Section 3, and the stability of the fixed point $\sqrt[n]{k}$ of WAM depends on the parameters n and p. Changing the parameter p of WAM we get a new map subclass, for example, for n=2 have the FDM. From these fact, we tried to detect numerically the optimal value of the parameter p, to minimize the ACT over the whole WAM family. For this we used a FORTRAN program to measure extensively the WAM MCT varying parameters (n,p) on a uniform grid of 500×500 points, for a radic and k. The result for k=2 is shown in Figure 4(a). The region in gray corresponds to unstable fixed point $\sqrt[n]{k}$ map WAM, as found in Section 1.3. The other colors seen in the graph are the regions of stability of the fixed point. For best visualization the MCT scale of this figure is truncated at a maximum value of 20, and higher values as inked light gray.

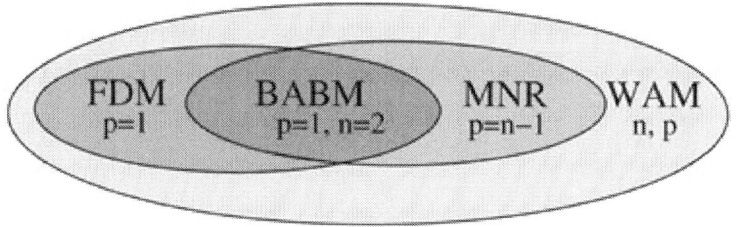

Figure 5: Hierarchy of the WAM subclasses.

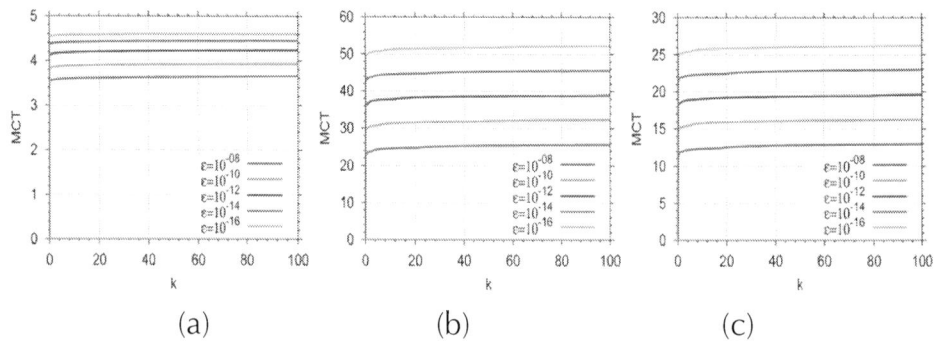

(a)　　　　　　　(b)　　　　　　　(c)

Figure 6: Numerical results of MCT for cubic roots (n=3) calculation with different precisions ε from 10^{-8} to 10^{-16} for (a) NRM; (b) FDM and (c) WAM with p=3.

We can see in Figure 4(a) that, as we approach the line that corresponds to the weighting term, NRM shows the minimum MCT over this line and therefore the most efficient of all studied maps is the NRM.

Summarizing the key information about NRM, FDM and WAM, with the numerical results for the MCT within double precision for these maps, for the parameters n=3 and $k \in [10^{-1}, 10^2]$, as shown in Table 2.

Analytical Convergence Time Model

The Lyapunov characteristic exponent for a unidimensional map $x_{i+1} = f(x_i)$ usually defined by

$\lambda(x_0) = \lim_{N \to \infty} \dfrac{1}{N} \sum_{i=0}^{(N-1)} \ln|f'(x_i)|$ can be approximate by $\lambda^* \approx \ln|f'(x^*)|$ since the derivative of the mapping function at the fixed point $f'(x^*)$ defines the rate of convergence of its time series, after discarded the transient.

For WAM, it is easy to show that $f'(x^*) = 1 - n/(p+1)$, so that this derivative assumes the value 1-n/2 when p=1 (FDM) and is zero only if p=n-1 (NRM or BABM, if n=2). In Figure 3(c) we observe that $f'(x^*) = 0$, for n=2 FDM reduces to NRM, and in Figure 3(d), $f'(x^*) = -1/2$ Using the original Lyapunov's idea, the characteristic exponent λ measures the average rate of convergence between two solutions separated by an initial distance ε_0, that is the case of time series dominated by a fixed point. For this orbits, the distance after i iterates is $\varepsilon_i = \varepsilon_0 e^{i\lambda}$ so that, if we assume that one orbit is initialized at $x_0 = x^*$ and other at $x_0 = x^* + 1$, i.e., the initial distance is unitary between orbits, we can use last equation to measure the error found in the second orbit, the root to be approximated. Within the standard double precision, the maximum error is of order 10^{-D}, where D is the number of decimal significants, typically $D = 15 \sim 16$ places.

Applying the natural logarithm to both sides of the above equation we have $i = \dfrac{\ln(\varepsilon_i)}{\lambda}$ for the number of iterations needed to reduce the error in the second orbit to ε. In the same manner we defined MCT, we define now the analytical convergence time (ACT), estimated by

$$ACT \approx \frac{-D\ln 10}{\ln\left|f'\left(x^*\right)\right|},$$

(1.7)

valid for any fixed point of a unidimensional map, where the approximated λ^* was used.

Applying this model to our more general map (WAM), we have

$$ACT(n, p) \approx \frac{-D\ln 10}{\ln\left|1 - n/(p+1)\right|}$$

(1.8)

of the parameters n and p. To double precision this approximated model function is plotted in Figure 4(b), that is remarkably very close to the numerical version plotted in Figure 4(a). Both figures uses the same color palette and truncated maximum, for better comparisons.

Table 2: MCT numerical results for NRM, FDM and WAM maps

Map	Estability	MCT (n = 3)
NRM	$\forall k,\ \forall n$	≈ 4.6
FDM	$n \in (0,4)$	≈ 52
WAM	$p > n/2 - 1$	≈ 26, for p=3

CONCLUSIONS

In the study of iterated maps to extract the real root of real numbers we have applied some common tools from nonlinear dynamics that allowed us to predict the fixed point of the studied maps associated with the nth root of k, and their stabilities could be analyzed in details.

We conclude, through the geometric argument used to recover the original analytical form of FDM, that both NRM and FDM can be reduced to averages between two terms, one linear and other nonlinear. From this observation, we generalize the original FDM idea to a new family of maps on which we add a new parameter p, whose value defines the stability of the map, the WAM. We show that FDM and NRM belong to this family of maps, being FDM recover when p=1 and NRM when p=n-1.

The mean convergence time (MCT) numerical results indicate that NRM is the most efficient subclass of the more general weighted average map (WAM) proposed in this work, as pointed out in Figure 4(a), over the line p=n-1. The analytical model for ACT is in complete agreement with the numerical results for WAM, the most general class of map studied. The model presented in Equation 1.7 is general, and can be adapted to any unidimensional map to study its fixed point attractor.

The main results of this work are obtained for $x \in R$, but their generalization is straightforward over the complex set C.

ACKNOWLEDGEMENTS

This work was partially supported by the Brazilian agency Conselho Nacional de Desenvolvimento Cientfico e Tecnológico—CNPq and Universidade do Estado de Santa Catarina—UDESC.

REFERENCES

1. Faber, X. and Voloch, J.F. (2011) On the Number of Places of Convergence for Newton's Method over Number Fields. Journal de Theorie des Nombres de Bordeaux, 23, 387-401.
2. Grau-Sánchez, M. and Daz-Barrero, J.L. (2011) A Technique to Composite a Modified Newton's Method for Solving Nonlinear Equations. ArXiv e-prints.
3. Pan, B., Cheng, P. and Xu, B. (2005) In-Plane Displacements Measurement by Gradient-Based Digital Image Correlation. SPIE Proceedings, 5852, 544-551.
4. Amin, A.M., Thakur, R., Madren, S., Chuang, H.-S., Thottethodi, M., Vijaykumar, T., Wereley, S.T. and Jacobson, S.C. (2013) Software-Programmable Continuous-Flow Multi-Purpose Lab-on-a-Chip. Microfluidics and Nanofluidics, 15, 647-659.http://dx.doi.org/10.1007/s10404-013-1180-2
5. Mungan, C.E. and Lipscombe, T.C. (2012) Babylonian Resistor Networks. European Journal of Physics, 33, 531. http://dx.doi.org/10.1088/0143-0807/33/3/531
6. Senthilpari, C., Mohamad, Z.I. and Kavitha, S. (2011) Proposed Low Power, High Speed Adder-Based 65-nm Square Root Circuit. Microelectronics Journal, 42, 445-451.http://dx.doi.org/10.1016/j.mejo.2010.10.015
7. Sun, T., Tsuda, S., Zauner, K.-P. and Morgan, H. (2010) On-Chip Electrical Impedance Tomography for Imaging Biological Cells. Biosensors and Bioelectronics, 25, 1109-1115.http://dx.doi.org/10.1016/j.bios.2009.09.036
8. Ausloos, M. and Dirickx, M. (2005) The Logistic Map and the Route to Chaos: From the Beginnings to Modern Applications. Springer, New York.
9. Eve, J. (1963) Starting Approximations for the Iterative Calculation of Square Roots. The Computer Journal, 6, 274- 276. http://dx.doi.org/10.1093/comjnl/6.3.274
10. Dellajustina, F.J. and Martins, L.C. (2014) The Hidden Geometry of the Babylonian Square Root Method. Accepted by Applied Mathematics, August.
11. Lyapunov, A.M. (1992) The General Problem of the Stability of Motion. International Journal of Control, 55, 531- 534. http://dx.doi.org/10.1080/00207179208934253
12. Schuster, H.G. and Just, W. (2005) Deterministic Chaos: An Introduction. 4th Edition, John Wiley & Sons, New York. http://dx.doi.org/10.1002/3527604804.

CITATION

Dias, C. , Dellajustina, F. and Martins, L. (2014) Two New Iterated Maps for Numerical Nth Root Evaluation Applied Mathematics, 5, 2974-2981. doi:10.4236/am.2014.519283.

Finite Type Transcendental Entire Functions Whose Buried Points Set Contains Unbounded Positive Real Interval

Feng Guo
Department of Mathematics, China University Mining & Technology, Beijing, China

2

ABSTRACT

Let $f_\mu(z) = z.e^{p(z)+\mu}$ with $P(z)$ being real coefficient polynomial and its leading coefficient be positive, $\mu \in \mathbb{R}^+$ when $P(z)$ and μ satisfy two certain conditions, buried point set of $f_\mu(z)$ contains unbounded positive real interval.

INTRODUCTION AND MAIN RESULT

Let $f(z)$ be an entire function on the complex plane $\mathbb{C}$. Define the iterated sequence $\{f^n\}$ of f as

$$f^0 = z, \quad f^{n+1}(z) = f \circ f^n(z) \quad (n = 0,1,2,\cdots)$$

$\mathbb{C}$ can be divided into two sets:

$$F(f) = \{z | \{f^n\} \text{ is normal at } z\}, \quad J(f) = \mathbb{C} \setminus F(f)$$

$F(f)$ is called Fatou set, which is open and contains at most countably many components. $J(f)$ is called Julia set, and it's closed and perfect. The fundamental theory of complex dynamical system can refer to [1] -[4] .

For an entire function f, let $S = \left\{ f \text{ is entire fuction} \middle| \sin g\left(f^{-1}\right) \text{is a finite set} \right\}$, with $\sin g\left(f^{-1}\right)$ be the set of singular value. If f is not a smooth covering map over any neighborhood of α, then α is a singular value. If $f \in S$, we call f is finite type entire function. The basic properties of the type entire function can refer to [5] .

I.N. Baker has first structured the transcendental entire function whose Julia set is $\mathbb{C}$ (See [6]):

Theorem A

For a certain real value μ, $f_{\mu} = ze^{z+\mu}$ has the whole complex plane for its Julia set.

Notice the set $\left\{ \mu \in \mathbb{R}^{+} : J\left(f_{\mu}(z)\right) = \mathbb{C}, f_{\mu} = ze^{z+\mu} \right\}$. Baker's result show that the set is nonempty; Jang, C.M. proved that this set contains infinitely many elements in [7]; Qiao, J. proved that the set is unbounded in [8] . What's more, Qiao has researched the buried sets in [9] , which contains unbounded positive real interval:

Theorem B

If $\mu \in [0, +\infty]$ for $f_{\mu} = ze^{z+\mu}$ and $J\left(f_{\mu}(z)\right) \neq \mathbb{C}$, then $\mathbb{R}^{+}$ belongs to the set of buried points.

Here we study the function $f_{\mu}(z) = z.e^{P(z)+\mu}$ with $P(z)$ is real coefficient polynomial and its leading coefficient is positive, expend the function in Theorem B:

Theorem 1 Let $f_\mu(z) = z.e^{P(z)+\mu}$, $P(z)$ is real coefficient polynomial and its leading coefficient is positive, the zeros of $ZP'(z)+1=0$ are $\{a_k | k = 1,\cdots,d\}$ which are real, unbounded positive real interval $L = (c,+\infty) \cap \mathbb{R}^+$ with $c = \max\limits_{1\geq k \leq d} \{a_k\}$. $J(f_\mu(z)) \neq \mathbb{C}$ and μ satisfy:

$$p(z) + \mu > 0 \ (z \in \mathbb{R}^+ \cup \{0\}) \tag{1}$$

$$x_\mu p'(x_\mu) \in (-\infty, -2) \cup (0, +\infty) \tag{2}$$

with x_μ is real zeros of $p(z) + \mu = 0$.

Then L belongs to the set of buried points set.

Remark: Qiao has given the example that satisfy the condition of Theorem B in [9] , then the example show that the function satisfy conditions in Theorem 1 is nonempty.

PROOF OF THEOREM 1

Lemma 1 Let $f(z)$ be an entire funcion of finite type. Then each Fatou component is eventually periodic, and $F(f)$ has only finitely many periodic compinents. They are attractive domains, superattractive domains, parabolic domains or Siegel discs.

Lemma 2 Let $f(z)$ be a transcendental entire function, D be a component of $F(f)$. If D is an attractive, a super attractive or a parabolic periodic domain, then the cycle of D contains at least one singularity of f^{-1}; if D is a Siegel disc, then the forward orbits of the singularities of f^{-1} are dense on ∂D.

Lemma 3 Let $f(z)$ be transcendental entire function of finite type. Then $f^n(z) \nrightarrow \infty (n \rightarrow \infty)$ for $z \in F(f)$.

Proof of Theorem 1: The singularities of f_μ^{-1} are 0 and $f_\mu(a_k) \in \mathbb{R}(1 \leq k \leq d)$, then

$\{f_\mu^n(a_k)\}_{n=1}^\infty \in \mathbb{R}(1 \leq k \leq d)$ if $J(f_\mu(z)) \neq \mathbb{C}$, so from Lemma 2 $F(f_\mu)$ have no Siegel disc and from Lemma 1,the periodic component of $F(f_\mu)$ only be attractive, superattractive or parabolic. $f_\mu(0) = 0$ and from $p(z) + \mu > 0 (z \in \mathbb{R}^+)$ we can have $f_\mu'(0) = e^{p(0)+\mu} > 1$, then 0 is a repelling fixed point, from Lemma 1 and 2, $F(f_\mu)$ has at most d cycles of periodic components $\{D_j^k\}_{j=1}^{P_k-1} (1 \leq k \leq d, p_k \in \mathbb{N})$:

$$\forall 1 \leq k \leq d, \quad f_\mu(D_j^k) = D_{j+1}^k \ (j = 0,1,2,\cdots, p_k - 2), \quad f_\mu(D_{p_k-1}^k) = D_0^k$$

such that $f_\mu(a_k) \in \bigcup_{i=0}^{p_k} D^k$ and there exist $\{x_k \in \mathbb{R}(1 \leq k \leq d)\}$ such that

$$* \quad f_\mu^{np_k+1}(a_k) \rightarrow x_k \ (n \rightarrow \infty)$$

Here we first proof $L = (c,\infty) \bigcap \mathbb{R}^+$ belong to $J(f_\mu)$. If there exist $t \in (c,\infty) \bigcap \mathbb{R}^+$ and $t \in F(f_\mu)$, then t is contained in a component of $F(f_\mu)$ and from above we have that there exist $m \in \mathbb{N}$, $a \in \{s_k | k = 1,\cdots,d\}$ a cycle of component $\{D_j\}_{j=1}^p$ with $p \in \{p_k | k = 1,\cdots,d\}$ and $x \in \{x_k | k = 1,\cdots,d\}$ such that $f_n^m(t)$ and $f_\mu(a)$ are in the same domain D_j, from * we have that $f_\mu^{m+np}(t) \to x_0 (n \to \infty)$. However, $p(z) + \mu > 0 (z \in \mathbb{R}^+)$ means $f_\mu(z) > z$ when $z \in \mathbb{R}^+$, $f_\mu'(z) > 0$ when $z \in (c,\infty) \bigcap \mathbb{R}^+$, then $\{f_\mu^n(z)\}_{n=1}^{\infty} \in L$ is montone increasing sequence, by the relation

$$f_\mu^{n+1}(z) = f_\mu^n(z)\exp\left(p\left(f_\mu^n(z)\right)+\mu\right)$$

we have $f_\mu^n(z) \to +\infty \left(z \in L \bigcap F(f_\mu); n \to \infty\right)$ which give a contradiction to Lemma 3.

Then we will proof that $L = (c,\infty) \bigcap \mathbb{R}^+$ belongs to the set of buried points. If there exist a point $a_0 \in (c,\infty) \bigcap \mathbb{R}^+$ and a_0 is on the boundary of a component of $F(f_\mu)$, from the discussion above, we know there exist some $N \in \mathbb{N}$, a cycle of component $\{D_j\}_{j=1}^p \in \{D_j^k\}_{j=1}^p (1 \le k \le d)$ with $p \in \{p_k | k = 1, \cdots, d\}$ such that when $n > N$,

$$f_\mu^n(a_0) \in \bigcup_{j=0}^{p-1} \partial D_j \quad (n = 1,2,3,\cdots)$$

and there exist $x_0 \in \{x_k \in \mathbb{R}(1 \le k \le d)\}$ and some $a_0 \in \{a_k | k = 1, \cdots, d\}$, $a_0 \in \{D_j\}_{j=1}^p$ such that $f_\mu^{np+1}(a) \to x_0 (n \to \infty)$.

Let $a_n = f_\mu^n(a_0)(n = 1,2,3\cdots)$, it's easy to have that $a_n \in \mathbb{R}^+$ and $a_{n+1} > a_n (n = 1,2,\cdots), a_n \to \infty(n \to \infty)$. Without the loss of generality, we can let $n > N$, $N, a_{np+j} \in \partial D (j = 0,1,2,\cdots,p-1)$.

Here we prove a_{np+j} $(n=1,2,3\cdots)$ are all in the same connected component of ∂D_j. If not, there exist a_{kp+j}, a_{sp+j} $(k<s)$ and two different component of ∂D_j called α_1 and α_2 such that

$$a_{kp+j} \in \alpha_1, \ a_{sp+j} \in \alpha_2$$

We can make curve ω, such that α_1 and α_2 belong to different components of $\mathbb{C} \backslash \omega$, then α_1 and α_2 belong to different components of $J(f_\mu)$, that gives a contradiction to $\left[a_{kp+j}, a_{sp+j}\right] \subset J(f_\mu)$.

Let $\delta_n \subset \partial D_j$ be an bounded continuum containing a_{np+j} and $a_{(n+1)p+j}$, we will prove that

$\left[a_{np+j}, a_{(n+1)p+j}\right] \subset \delta_n$. If not, δ_n and $\left[a_{np+j}, a_{(n+1)p+j}\right]$ can form a bounded domain and $J(f_\mu)$ have no interior point, therefore $J(f_\mu)$ have to have a bounded domain, notice that $\{a_n\}_{n=1}^{\infty}$, let $\lim_{n\to\infty} a_n = A$, due to $a_{n+1} = a_n e^{p(a_n)+\mu}$, then $A = Ae^{p(A)+\mu}$, notice that $a_n, A \in \mathbb{R}^+$ and $p(z)+\mu>0$ when $z \in \mathbb{R}^+$ therefore

$A = 0, \infty$, from $a_n \in \mathbb{R}^+$ we have $A = +\infty$. That means D_j is a unbounded component, however any component of $F(f_\mu)$ have to turn into cycle $\left(D_j\right)_{j=0}^{p-1}$ from Lemma 1, therefore the components of $F(f_\mu)$ are all unbounded, its contradiction. What's more, we can have that

$$\bigcup_{n\geq N}\left[a_{np+j}, a_{(n+1)p+j}\right] \subset \partial D_j \quad (j=1,2,3,\cdots,p-1)$$

$$\left[a_{Np+j}, \infty\right] \subset \partial D_j \quad (j=1,2,3,\cdots,p-1)$$

that means $\left[a_{N_0},\infty\right]$ is the common boundary of $D_0,D_1,\cdots,D_{p-1}$ with $N_0=Np+p-1$. The common boundary is at most of two domains, therefore $p\le 2$. Here we divide two cases to discuss:

Case 1: $p=1$. $f_\mu(a)\in\bigcup_{j=0}^{p-1}D_j$ is $f_\mu(a)\in D_0$. Considering

$$f_\mu^{n+1}(a)=f_\mu^n(a)e^{p\left(f_\mu^n(a)\right)+\mu}$$

and * we have

$$x_0=\lim_{n\to\infty}f_\mu^{n+1}(a)=0 \text{ or } x_0 \text{ are the zeros of } p(z)+\mu=2k\pi i \ (k=0,\pm 1,\pm 2,\cdots)$$

Notice $a\in\mathbb{R}$, then $x_0\in\mathbb{R}$, we only need to consider 0 and the real zeros of $p(z)+\mu=0$, from Lemma 2 x_0 is an attractive, super attractive, or rational indifferent fixed point, but from conditions (1) and (2) in Theorem 1, 0 and x_0 are repelling fixed point, it's a contradiction.

Case 2: $p=2$. Without the loss of generality, we take D_0 be the component above $\left[a_{N_0},+\infty\right)$ and D_1 be the component under $\left[a_{N_0},+\infty\right)$, for $r\in\left[a_{N_0},+\infty\right)$ with r is large enough, take a sequence $\{z_n\}_{n=1}^\infty\in D_0\cap\{z|\mathrm{Im}\,z>0\}$ such that $z_n\to r(n\to\infty)$.

Let $z_n=r_ne^{i\theta_n}$ with $r_n>0,\theta_n\in\left(0,\dfrac{\pi}{2}\right)$, then

$$f_\mu(z_n)=r_ne^{i\theta_n}e^{p(r_ni\theta_n)+\mu}$$

We can suppose that $p(z)=c_nz^d+\cdots+c_1z+c_0$, then we have

$$\operatorname{Im} p\left(r_n e^{i\theta_n}\right) = c_n r_n^d \sin\left(\theta_n d\right) + \cdots + c_1 r_n \sin\left(\theta_n\right) > \left(c_n r_n^d + \cdots + c_1 r_n\right)\sin\theta_n = p\left(r_n\right)\sin\theta_n$$

Notice that $p(z) + \mu > 0\left(z \in \mathbb{R}^+ \bigcup\{0\}\right)$ and we can easily deduce that $mp\left(r_n e^{i\theta_n}\right) > p(r_n)\sin\theta_n > 0$ and $f_\mu\left(z_n\right)$ belong to the above half plane when n and r is large enough, but it contradicts that $f_\mu\left(z_n\right) \in D_1$. The proof is complete.

ACKNOWLEDGEMENTS

Author thanks professor Jianyong Qiao and Yuhua Li for their discussions and suggestions.

REFERENCES

1. Eremenko, A.E. and Lyubich, M.Yu (.1990) the Dynamics of Analytic Transformations. Leningrad Mathematical Journal, 1, 563.
2. Beardon, A.F. (1991) Iteration of Rational Functions. Springer, Berlin.http://dx.doi.org/10.1007/978-1-4612-4422-6.
3. Milnor, J. (2006) Dynamics in One Complex Variable. 3rd Edition, Princeton University Press, Princeton and Oxford.
4. Qiao, J. (2010) Complex Dynamics on Renormalization Transformations. Science Press, Beijing. (in Chinese).
5. Morosawa, S., Nishimura, Y., Taniguchi, M. and Ueda, T. (2000) Holomorphic Dynamics, Cambridge University Press.
6. Baker, I.N. (1970) Limit Functions and Sets of Non-Normality in Iteration Theory. Annales Academiae Scientiarum Fennicae. Series a 1, Mathematica, 467, 1-11.
7. Jang, C.M. (1992) Julia Set of the Function z exp (z + μ). Tohoku Mathematical Journal, 44, 271-277.
8. Qiao, J. (1994) The Set of the Mapping $z \rightarrow$ exp (z + μ). Chinese Science Bulletin, 39, 529.
9. Qiao, J. (1995) the Buried Points on the Julia Sets of Rational and Entire Functions. Science in China Series A, 38, 1409-1419.

CITAION

Guo, F. (2014) Finite Type Transcendental Entire Functions Whose Buried Points Set Contains Unbounded Positive Real Interval. Advances in Pure Mathematics, 4, 209-212. doi: 10.4236/apm.2014.45027.

Coherence Modified for Sensitivity to Relative Phase of Real Band-Limited Time Series

William Menke
Lamont-Doherty Earth Observatory of
Columbia University, Palisades, NY, USA

ABSTRACT

As is well known, coherence does not distinguish the relative phase of a pair of real, sinusoidal time series; the coherence between them is always unity. This behavior can limit the applicability of coherence analysis in the special case where the time series are band-limited (nearly-monoch- romatic) and where sensitivity to phase differences is advantageous. We propose a simple modification to the usual formula for coherence in which the cross-spectrum is replaced by its real part. The resulting quantity behaves similarly to coherence, except that it is sensitive to relative phase when the signals being compared are strongly band-limited. Furthermore, it has a useful interpretation in terms of the zero-lag cross-correlation of real band-passed versions of the time series.

INTRODUCTION

In this paper, we examine the well-known formula for the frequency-dependent coherence C^2 between two time series and argue that it is not well-suited for quantifying the similarity of band-limited data. Using a time domain-based analysis, we identify a critical step in the development of the traditional algorithm, which we show is inappropriate in the band-limited case, and propose an alternative that leads to the definition of a new quantity S^2, which while having a definition similar to C^2, is better behaved. We then use both synthetic tests and analytic methods to elucidate the behavior of S^2, and show that it is a viable alternative to C^2. Our belief is that the choice of time series analysis technique should be guided by the properties of the data; one analyzes time series in a way designed to best extract knowledge from them. One should always be willing to adapt an analysis method to achieve this goal.

The issue considered here is how best to quantify the similarity between time series that are 1) real (as contrasted to complex) and 2) band-limited (in the sense of being nearly monochromatic). Such time series constitute important special cases because most natural phenomena are described using real numbers and many are dominated by a single period of oscillation. For example, the daily period often contributes strongly to physiological and meteorological signals, the annual period to environmental and climatic signals, the precessional period (25.7 ka) [1] to sedimentary and paleontological signals, and so forth. Furthermore, commonly-used techniques such as multiple window coherence analysis [2] , where two long time-series are divided into a sequence of shorter pairs before coherence analysis is performed, may accentuate the degree to which a single period of oscillation dominates the signal.

An important property of nearly-monochromatic signals is their relative phase. Whether two time series that are in-phase (as in Figure 1(a)) or out-of-phase (as in Figure 1(b)) may be important, for example, from the perspective of an analyst trying to unravel the dynamics of the underlying causative processes.

Traditional coherence analysis [3] has very limited application in this case, because of the well-known insensitivity of coherence to relative phase. The coherence of two sinusoidal time series of the same period is always unity, irrespective of their relative phase. Simply put, coherence does not distinguish a sine from a cosine. Given the general usefulness of coherence in other settings, it is well to ask why it "fails" in this special case and whether it can be modified to produce what may, in some circumstances, be a more useful measure of similarity.

When asking why any quantity encountered in time series analysis, such as coherence, behaves in a certain way, one must contend with the fact that most, if not all, such quantities can be derived from several different perspectives. Any answer will probably make sense only from one of these points of view. Consider, for example, the estimated mean of a time series. This deceptively simple quantity can be understood, alternately, as arising through the minimizing of error (a deterministic derivation) [4] or through the maximizing of likelihood (a probabilistic derivation) [5] or through the maximization of importance (an informational derivation) [6] to name just a few. The answer to a question concerning the estimated mean, say for example, whether it should always be bounded by the smallest and largest datum, will necessarily refer to one of these perspectives. The same is true for coherence. We adopt here a deterministic perspective:

The coherence between two time series, at frequency, ω_0, is closely-related to the zero-lag cross-correlation of band-passed versions of those time series, where the band-pass filter is one-sided and has center-frequency, ω_0. In fact, the former is merely a normalized and squared version of the latter.

This is but one perspective among many, but one we find helpful because it brings out a relationship to the cross-correlation, another quantity useful in assessing the similarity between two time series. Cross-correlation is defined in the time-domain, as contrasted to coherence, which is defined in the frequency-domain, so the link provides complimentary information.

The appearance of a one-sided filter (Figure 2(a)), may seem counter-intuitive, because such filters are almost never used in practice, or at least not when the data are real, for they turn a real-time series into a complex one. All the band-pass filters that an analyst would commonly use are two-sided (Figure 2(b)), and so have real output. The reason for its appearance here is that the usual definition of coherence is completely general. It does not presume that the signals being compared are real, and so builds in the possibility that negative and positive

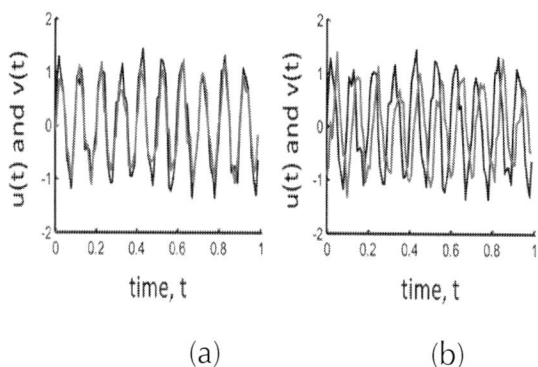

(a)   (b)

Figure 1: (a) Two nearly-monochromatic time series (black and red curves) with the relative

phase, $\varphi = 0$; (b) Same as (a), but with relative phase, $\varphi = \pi / 2$.

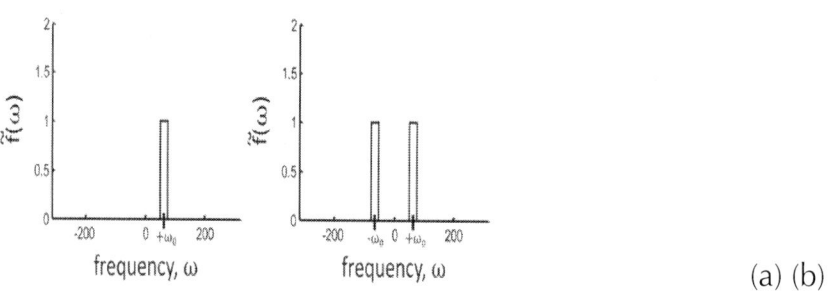

(a) (b)

Figure 2: (a) Fourier transform of a one-sided band-pass filter, consisting of a single boxcar function centered at frequency, ω_0, and with width, $2\Delta\omega$; (b) Fourier transform of a two-sided band-pass filter, also centered at frequency, ω_0.

frequency components of the time series behave completely differently from one another. This is contrast to real-time series, where they are complex conjugate pairs. However, in being general, it cannot exploit an important property of real signals: that sines and cosines are distinguishable from one another. As we show below, substituting a two-sided filter produces a version of coherence that distinguishes sines from cosines; that is, one that is sensitive to the relative phase of band-limited signals.

COHERENCE-LIKE MEASURE OF SIMILARITY BASED OF CROSS-CORRELATION

The problem we consider is how to quantify the similarity of two real, transient time series, u(t)and v(t), in the vicinity of a specified frequency, ω_0. The strategy we adopt is to band-pass filter these time series and then to compute their zero-lag cross-correlation. The filter selects out frequencies near ω_0 and the cross-correlation quantifies similarity, since it attains its largest value when u(t) = v(t) (ignoring, for the moment, the issue of normalization). We denote the filtered time series as, $f(t)*u(t)$ and $f(t)*v(t)$ where the symbol $*$ denotes convolution. We require the filtered time series to be purely real, so that the filter, f(t) has a two-sided Fourier transform with the symmetry, $\tilde{f}(\omega) = \tilde{f}^*(-\omega)$, where the tilde denotes Fourier transformation and the asterisk denotes complex conjugation. We choose a filter with a purely-real Fourier transform, built from two unit-am- plitude boxcar functions, one centered at $-\omega_0$ and the other at $+\omega_0$, each of width $2\Delta\omega$. This filter does not affect Fourier components within the pass-band and completely rejects those outside of it.

The convolution, g(t), and cross-correlation,c(t) , of two real time series are defined as [7] (their pages 24 and 46):

$$g(t) = u * v = \int_{-\infty}^{+\infty} u(\tau)v(t-\tau)\,d\tau$$

(1a)

$$c(t) = u * v = \int_{-\infty}^{+\infty} u(\tau)v(t+\tau)\,d\tau$$

(1b)

Note that at zero-lag, cross-correlation is just the area beneath the product of the two time series:

$$c(t = 0) = (u * v)_{t=0} = \int_{-\infty}^{+\infty} u(\tau) v(\tau) \, d\tau$$

(2)

Note also that definition of the convolution and cross-correlation in (1a), (1b) differ only by a sign of τ in the $v(t \pm \tau)$ term. The substitution, $\tau' = -\tau$, leads to the very useful, well-known identity, $a(-t) * b(t) a(t) * b(t)$ ([7], their page 47). Applying this identify, we find that the cross-correlation of the filtered time series is:

$$c(t, \omega_0, \Delta\omega) = \{ f(t, \omega_0, \Delta\omega) * u(t) \} * \{ f(t, \omega_0, \Delta\omega) * v(t) \} = f(-t, \omega_0, \Delta\omega) * f(t, \omega_0, \Delta\omega) * u(-t) * v(t)$$

(3)

At zero lag, the cross-correlation is proportional the integral of its Fourier transform, $\tilde{c}(\omega)$:

$$c(t = 0) = \frac{1}{2\pi} \int_{-\infty}^{+\infty} \tilde{c}(\omega) \exp(i\omega t) \Big|_{t=0} \, d\omega = \frac{1}{2\pi} \int_{-\infty}^{+\infty} \tilde{c}(\omega) \, d\omega$$

(4)

Inserting (3) into (4) and using the rule that the Fourier transform of a convolution is the product of the transforms ([7] , page 115) and the rule that the transform of $a^*(-t)$ is $\tilde{a}^*(\omega)$ (see Appendix) yields:

$$c(t = 0, \omega_0, \Delta\omega) = \frac{1}{2\pi} \int_{-\infty}^{+\infty} \tilde{f}(\omega, \omega_0, \Delta\omega) \tilde{f}(\omega, \omega_0, \Delta\omega) \tilde{u}^*(\omega) \tilde{v}(\omega) \, d\omega$$

$$\approx \frac{1}{2\pi} \int_{-\omega_0 - \Delta\omega}^{-\omega_0 + \Delta\omega} \tilde{u}^*(\omega) \tilde{v}(\omega) \, d\omega + \frac{1}{2\pi} \int_{+\omega_0 - \Delta\omega}^{+\omega_0 + \Delta\omega} \tilde{u}^*(\omega) \tilde{v}(\omega) \, d\omega$$

$$= \frac{1}{\pi} \int_{\omega_0 - \Delta\omega}^{\omega_0 + \Delta\omega} \mathrm{Re}\{ \tilde{u}^*(\omega) \tilde{v}(\omega) \} \, d\omega = \frac{2\Delta\omega}{\pi} \overline{\mathrm{Re}\{ \tilde{u}^*(\omega_0) \tilde{v}(\omega_0) \}}.$$

(5)

Here $\bar{\tilde{a}}(\omega_0)$ denotes the mean value of $a(\omega)$ in the frequency band, $\omega_0 \pm \Delta\omega$. Note that $c(t = 0, \omega_0, \Delta\omega)$

is defined for $\omega_0 > 0$, only. The quantity $\tilde{u}^*(\omega)\tilde{v}(\omega)$ is the cross-spectrum. Thus, the zero-lag cross-correla- tion of the real band-pass filtered time series depends upon the average value of the real part of their cross-spec- trum in the filter's pass-band. The amplitude of $c(t = 0, \omega_0, \Delta\omega)$ depends on the amplitude of two time series, as well as upon their de- gree of similarity. We remove this dependence by normalizing by the energy E_u and E_v in the two time series, defined as:

$$E_u = \left(u * u\right)_{t=0} = \int_{-\infty}^{+\infty} u^2(\tau)\,d\tau \text{ and } E_v = \left(v * v\right)_{t=0} = \int_{-\infty}^{+\infty} v^2(\tau)\,d\tau \tag{6}$$

The normalized measure of similarity, say S, is:

$$S = \frac{c\left(t = 0, \omega_0, \Delta\omega\right)}{E_u^{1/2} E_v^{1/2}} = \frac{\overline{\mathrm{Re}\left\{\tilde{u}\left(\omega_0\right)\tilde{v}\left(\omega_0\right)\right\}}}{\left|\tilde{u}\left(\omega_0\right)\right|\left|\tilde{v}\left(\omega_0\right)\right|} \text{ with } 0 \le \omega_0 < +\infty \tag{7}$$

Note that the quantity, S^2, which we nickname here similarity, varies between zero and unity. It has almost exactly the functional form of the quantity called coherence, except for the taking of the real part. The imaginary part cancelled from (5) precisely because the time series are real and the filter is two-sided.

COHERENCE RELATED TO ZERO-LAG CROSS-CORRELATION

As asserted in the Introduction, the usual formula for coherence can be obtained simply by switching to a one-sided filter, a single unit step func- tion of width $2\Delta\omega$ centered at frequency ω_0 (where $-\infty < \omega_0 < +\infty$). The filtered time series $f * u$ and $f * u$ are complex, so that one must define a cross-correlation appropriate for complex signals; that is, replace

$u(\tau)$ with $u^*(\tau)$ in (1b). These modifications lead to a version of (7) that is exactly the usual formula for the coherence:

$$C^2\left(\omega_0,\Delta\omega\right)=\frac{c^2\left(t=0,\omega_0,\Delta\omega\right)}{E_u E_v}=\frac{\left|\tilde{u}^*\left(\omega_0\right)\tilde{v}^*\left(\omega_0\right)\right|^2}{\left|\tilde{u}\left(\omega_0\right)\right|^2\left|\tilde{v}\left(\omega_0\right)\right|^2} \quad \text{with} \quad -\infty<\omega_0<+\infty$$

(8)

As an aside, we note that our derivations of $C^2\left(\omega_0\right)$ and $S^2\left(\omega_0\right)$ hide an inconsistency in the interpretation of $\left|\tilde{u}(\omega_0)\right|^2$ as the power in the time series $u(t)$ at frequency, ω_0. It represents power for a complex time series but only half the power for a real one, owing to the different intervals over which frequency, ω_0, is defined. This factor of two compensates for the apparent loss of power when the real part is taken in (5).

SIMILARITY AND COHERENCE OF REAL BAND-LIMITED SIGNALS

Suppose that time series $u(t)$ and $v(t)$ are monochromatic, with equal frequency, ω_0, but with different amplitudes, u_0 and u_0, and relative phase, φ:

$$u(t)=u_0\sin\left(\omega_0 t\right) \quad \text{and} \quad v(t)=v_0\sin\left(\omega_0 t-\varphi\right)$$

(9)

The similarity, $S^2\left(\omega_0\right)$, is most easily calculated using its time-domain definition. Taking, without loss of generality, the window of observation to be $0<\tau<2\pi$, we have:

Coherence Modified for Sensitivity to Relative Phase of Real Band

$$E_u = u_0^2 \int_0^{2\pi} \sin^2(\omega_0 \tau) d\tau = \frac{\omega_0 u_0^2}{2} \quad \text{and} \quad E_v = v_0^2 \int_0^{2\pi} \sin^2(\omega_0 \tau - \varphi) d\tau = \frac{\omega_0 v_0^2}{2}$$

$$\text{and} \quad c(t=0) = u_0 v_0 \int_0^{2\pi} \sin(\omega_0 t) \sin(\omega_0 t - \varphi) d\tau = \frac{\omega_0 u_0 v_0}{2} \cos \frac{\omega_0 u_0 v_0}{2} \cos(\varphi)$$

$$\text{so} \quad S^2 = \frac{c^2(t=0)}{E_u E_v} = \cos^2(\varphi).$$

$$(10)$$

Thus, S^2 is unity when the two sinusoids are in-phase $(\varphi = 0)$ and declines monotonically to zero when they are out-of-phase $(\varphi = \pi/2)$.

The coherence, $C^2(\omega_0)$, is calculated by recognizing that a sine function is built up of two complex exponentials of frequency $+\omega_0$ and $-\omega_0$ and that the one-sided filter selects only the one with positive frequency:

$$f(t) * u(t) = U_0 \exp(i\omega_0 t) \quad \text{and} \quad f(t) * v(t) = V_0 \exp(i\omega_0 t)$$

$$\text{with} \quad U_0 = \frac{u_0}{2i} \quad \text{and} \quad V_0 = \left\{ \frac{u_0}{2i} \cos(\varphi) - \frac{v_0}{2i} \sin(\varphi) \right\}.$$

$$(11)$$

We then find:

$$E_u = U_0^2 \int_0^{2\pi} \exp(-i\omega_0 t) \exp(i\omega_0 t) d\tau = 2\pi U_0^2 \quad \text{and} \quad E_v = 2\pi V_0^2$$

$$\text{and} \quad c(t=0) = 2\pi U_0 V_0 \quad \text{so} \quad C^2 = \frac{c^2(t=0)}{E_u E_v} = 1,$$

$$(12)$$

This is the well-known result that the coherence, C^2, is unity irrespective of the relative phase of the two sinusoids. This behavior is a consequence of the one-sided filter, which turns both $\sin(\omega_0 t)$ and $\cos(\omega_0 t)$ into functions proportional to the same complex exponential, $\exp(i\omega_0 t)$.

EXAMPLES

We consider the example of a sequence of nearly-monochromatic wavelets, formed by taking the product of a phase-shifted sinusoid of frequency, ω_0, and a normal envelope function of half-width, σ:

$$\sin\left(\omega_0 t - \varphi\right)\exp\left(\left(t - t_0\right)^2 / \left(2\sigma^2\right)\right) \tag{13}$$

and then by adding a small amount of uncorrelated random noise. Figures 3(a)-(c) illustrate pairs of these wavelets with different phase relationships. Note that the wavelets are not merely time-shifted versions of one another, since the position of the zeros crossings of the sinusoid (parameterized by φ) can and the position of the center of the envelope (parameterized by t_0) can be independently varied. One might imagine a time series analysis scenario where u(t) represents the external forcing applied to some dynamical system, and v(t) represents the response. In such a context, the distinction between these different wavelet shapes is important, say for detecting whether or not some anticipated interaction has occurred. In this case, the similarity, $S^2(\omega_0)$ (red curves in Figure 3(d), Figure 3(c)) is a more useful quantity than the coherence, $C^2(\omega_0)$ (black curves), since it varies strongly with the phase-relationships, whereas coherence does not.

We have not performed an exhaustive analysis of the differences between $S^2(\omega_0)$ and $C^2(\omega_0)$, when they are applied to broad-band signals. The key difference is the effect of the taking of the real part:

$$S^2 \propto \left(\overline{\tilde{u}_R \tilde{v}_R} + \overline{\tilde{u}_I \tilde{v}_I}\right)^2,$$

$$C^2 \propto \left(\overline{\tilde{u}_R \tilde{v}_R} + \overline{\tilde{u}_I \tilde{v}_I}\right)^2 + \left(\overline{\tilde{u}_R \tilde{v}_I} - \overline{\tilde{u}_I \tilde{v}_R}\right)^2. \tag{14}$$

where the Fourier transforms are written in terms of their real and imaginary parts, $\tilde{u} = \tilde{u}_R + i\tilde{u}_I$ and $\tilde{v} = \tilde{v}_R + i\tilde{v}_I$. Since S^2 and C^2 differ by a manifestly positive amount, we are guaranteed that $C^2 \geq S^2$. However, without further specification of the behavior or $\tilde{u}$ and $\tilde{v}$, no further characterization is possible. In the special case where both time series contain a common function w(t), so that $u(t) = w(t) + x(t)$ and $v(t) = w(t) + y(t)$ and where w(t), x(t) and y(t) are all broad-band, we find:

$$S \propto \left(\overline{\tilde{w}_R \tilde{w}_R} + \overline{\tilde{w}_I \tilde{w}_I} \right) + \text{crossterms like } \overline{\tilde{x}_R \tilde{y}_R},$$

$$C \propto \left(\overline{\tilde{w}_R \tilde{w}_R} + \overline{\tilde{w}_I \tilde{w}_I} \right) + \text{crossterms like } \overline{\tilde{x}_R \tilde{y}_I}.$$

(15)

We might expect in the case that $C^2 \approx S^2$, since the cross-terms are averages of functions that oscillate around zero and therefore likely to be small. Numerical tests (Figure 4) support this idea, at least for non-transient broad-band time series with a moderate degree of correlation.

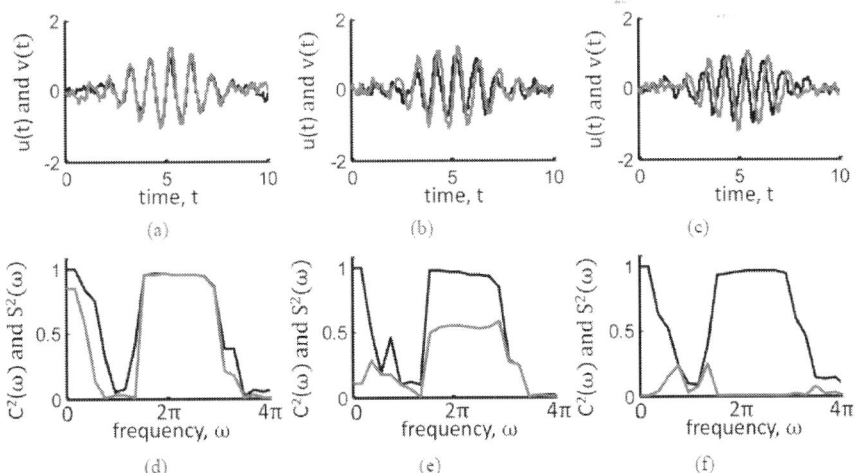

Figure 3: (a)-(c) Sequence of three pairs of nearly-monochromatic time series with frequency $\omega_0 \approx 2\pi$ and with relative phase of, $\varphi = 0$, $\varphi = \pi / 4$. And $\varphi = \pi / 2$, respectively; (d)-(f) Corresponding coherence, $C^2(\omega)$, and similarity (black curve) and $S^2(\omega)$ (red curve). Note that $C^2(\omega)$, is approximately unity for all three cases, whereas $S^2(\omega)$ decreases as the relative phase increases. In this example $\Delta\omega$ is set to $\omega_0 / 4$.

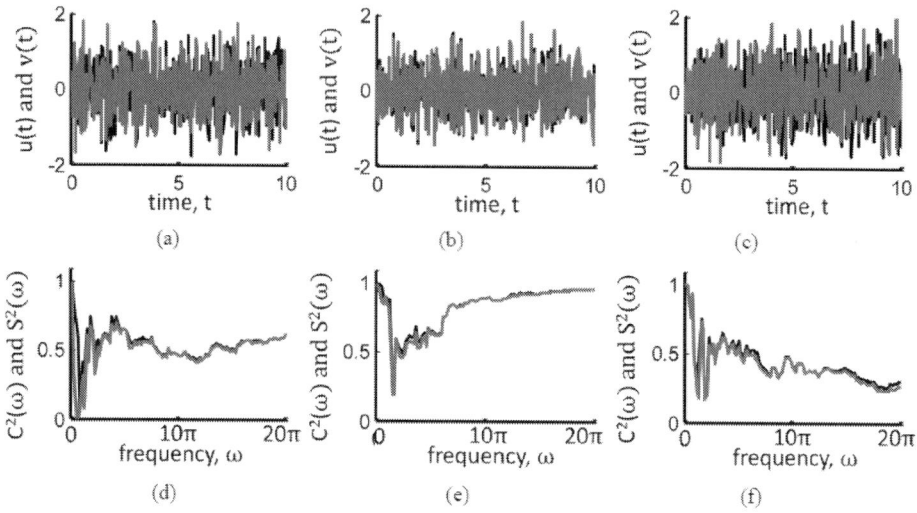

Figure 4: (a)-(c) Sequence of three pairs of broad-band time series with an approximately: (a) frequency-independent coherence, (b) coherence that increases with frequency, and (c) coherence that decreases with frequency; (d)-(f) Corresponding coherence, $C^2(\omega)$ (black curve), and similarity and $S^2(\omega)$ (red curve), which are approximately equal, although also obeying the rule $C^2 \geq S^2$. In this example $\Delta\omega$ is set to $\omega_0 / 4$.

CONCLUSION

In summary, we recommend this simple modification of coherence in cases where the time series that are being compared are narrow-band and where phase relationships between them are considered important. For pure sinusoids differing by phase, φ, it obeys the rule $S^2 = \cos^2\varphi$; that is, similarity monotonically decreases from unity, when $\varphi = 0$, and to zero, when $\varphi = \pi / 2$. In other respects, it behaves very similarly to coherence. Finally, it has a very intuitive time-domain interpretation: $S(\omega_0)$ gives you exactly what you would get if you normalized each time series by the square-root of its energy, band-pass filtered each with a two-sided boxcar filter centered around frequency, ω_0, and computed their zero-lag cross-correlation.

Appendix

The rule that the transform of $a^*(-t)$ **is** $\tilde{a}^*(\omega)$ is well-known, but we derive it here for completeness. The Fourier transform of $a^*(-t)$ is:

$$\int_{-\infty}^{+\infty} a^*(-t)\exp[-i\omega t]\,dt = \left[\int_{-\infty}^{+\infty} a(-t)\exp[i\omega t]\,dt\right]^* = \left[-\int_{-\infty}^{+\infty} a(\tau)\exp[-i\omega\tau]\,d\tau\right]^*$$

$$= \left[\int_{-\infty}^{+\infty} a(\tau)\exp[-i\omega\tau]\,d\tau\right]^* = \tilde{a}^*(\omega).$$

$$(A1)$$

Here we have utilized the substitution $t = -\tau$.

REFERENCES

1. Olsen, P.E. and Kent, D.V. (1996) Milankovitch Climate Forcing in the Tropics of Pangea during the Late Triassic. Paleooceanography, Paleoclimatology, Paleoecology, 122, 1-26.http://dx.doi.org/10.1016/0031-0182(95)00171-9
2. Bendat, J.S. and Piersol, A.G. (2010) Random Data, Anaysis and Measurement Procedures. John Wiley and Sons, New York. http://dx.doi.org/10.1002/9781118032428
3. Sandberg, J. and Hansson, M. (2006) Cohernece Estimation between EEG Signals Using Multiple Window Time- Frequency Analysis Compared to Gaussian Kernels. Proceedings of the 14th European Signal Processing Conference, Florence, 4-8 September 2006.http://www.eurasip.org/Proceedings/Eusipco/Eusipco2006/papers/1568981924.pdf
4. Lawson, C.L. and Hanson, R.J. (1974) Solving Least Squares Problems. Prentice-Hall, New York.
5. Millar, R.B. (2011) Maximum Likelihood Estimation and Inference: With Examples in R, SAS and ADMB. John Wiley and Sons, New York.http://dx.doi.org/10.1002/9780470094846
6. Menke, W. (2012) Geophysical Data Analysis: Discrete Inverse Theory. MATLAB Edition, Elsevier, Amsterdam.
7. Bracewell, R.N. (2000) The Fourier Transform and Its Applications. 3rd Edition, McGraw-Hill, New York.

CITATION

Menke, W. (2014) Coherence Modified for Sensitivity to Relative Phase of Real Band-Limited Time Series. Applied Mathematics, 5, 2739-2745. doi: 10.4236/am.2014.517261.

Multiple Factorial Analysis of Symbolic Data

Barnabé Tang Ahanda, Jean Gérard Aghoukeng Jiofack, Romain Germain Nzangué, and Gilbert Hapi Mbiakop
Department de Mathematics, Faculty of Sciences, University of Yaounde 1, Yaounde, Cameroon

ABSTRACT

This document presents an extension of the multiple factorial analysis to symbolic data and especially to space data. The analysis makes use of the characteristic coding method to obtain active individuals and the reconstitutive coding method for additional individuals in order to conserve the variability of assertion objects. Traditional analysis methods of the main components are applied to coded objects. Certain interpretation aids are presented after the coding process. This method was applied to poverty data.

INTRODUCTION

Although the traditional analysis of data is based on a complete theory, it poses restrictive conditions for the resolution of problems: it requires that variables be single-valued (the value of a variable for an object is a unique and specific).

Recent evolutions in the information system have made it possible to find increasing and more complex data which need a more in-depth formalization than the one indicated in the usual rectangular table. The later does not take into account the possible variability of description as well as the uncertainty concerning certain variables or certain objects. A new formalism called symbolic objects has been developed to represent complex data like concepts, skills (see [1, 2]). These objects concern complex data and provide new skills at the exit in the form of symbolic objects.

The techniques of factorial analysis (see [3,4]) which are data reduction and representation methods, are well known and well used. Certain stakeholders have extended these methods to symbolic objects by using the coding and decoding techniques (read [5,6]). The issue of multiple factorial analysis (MFA) is of particular interest to us (see [7,8]). A priori, this concerns the case where there is a group structure on variables. Its problem is enhanced by the characterization of groups and the search for a typology of groups of variables. To our knowledge, this method is not yet applied to symbolic objects.

In this article, we extend the MFA to a symbolic data base whose variables are structured into groups. It is a generalization of the MFA basic principle on the one hand, and the principle components analysis (PCA) techniques on the other hand (see [9]). We conclude this document with an application of the method presented to poverty data.

SYMBOLIC DATA BASE

General Context

A symbolic data base (SDB) is considered as represented in the Table 1 and made up of triplet B= {Y, C, A} with C={C_1, C_2, ..., C_I} representing all the I objects, Y={$y_1, y_2, ..., y_J$} and A= {a^1, a^2, ..., a^I} representing respectively the set of the J variables and the assertion objects used to describe I objects. We suppose that the SDB is associated with a measured space () an σ – algebra C on C, O= {$O_1, O_2, ..., O_J$}

where

O_j is the space of observation of the variable y_j, j=1, 2, ..., J.

Multiple Factorial Analysis of Symbolic Data

This space includes an tribe $_j$. Let $(X_{ci, j})_{j=1, 2, \ldots J}$ be a set of measurable functions defined on deriving their value from the measurable spaces $(o_{j'j})$ and used to describe all the c_i objects. For I from 1 to I,

$$a^i = \bigwedge_{j=1}^{J} \left[y_j = X_{a^i, j} \right]$$ a^i is a Boolean assertion so that from j from 1 to$J$, there is a measurable part $V_{a^i, j} \in _j$ so that we can have

$$X_{a^i, j} = X_{c_i, j} \text{ and } V_{a^i, j} = X_{a^i, j} ().$$

In practice, it is written

$$a^i = \bigwedge_{j=1}^{J} \left[y_j = V_{a^i, j} \right]$$

(1)

Table 1: Symbolic data base

	Y_1	Y_j	Y_J
C_1	$\left[y_1 = X_{a^1, 1} \right]$	$\left[y_j = X_{a^1, j} \right]$	$\left[y_J = X_{a^1, J} \right]$
C_i	$\left[y_1 = X_{a^i, 1} \right]$	$\left[y_j = X_{a^i, j} \right]$	$\left[y_J = X_{a^i, J} \right]$
C_t	$\left[y_1 = X_{a^i, 1} \right]$	$\left[y_j = X_{a^i, j} \right]$	$\left[y_J = X_{a^i, J} \right]$

The observation space is therefore the product space

$$\Theta_j = \prod_{j=1}^{J} o_j$$ This space includes an product tribe

$$\Xi_j = \prod_{j=1}^{J} \mathcal{O}_j$$ To proceed, it is supposed that the descriptors are quantitative, i.e that O_j is a part of $\mathbb{R}$.

Thus, it is generalized that

$$X_{a^i, j}(\Omega) = V_{a^i, j} = \left[\underline{x_{ij}}, \overline{x_{ij}} \right]$$

where $x_{ij}, \overline{x_{ij}}$ represent respectively the minimum and maximum values per $\overline{a}^i$ object for the y_j variable. Consequently, we shall treat the Boolean assertions whose SDB is indicated in the following Table 2.

Group Structure

We shall assume that the J variables are divided into K groups with J_k as the number of k group variables; p_i the weight associated with the assertion object a_i and m_ℓ the weight of the y_ℓ variable. We shall designate I, K, J_k as the set and its cardinal. Let us consider $B_k = \{Y_k, C, A_k\}$ as the symbolic data base derived from B but reduced from the k group variables denoted Y_k. For i from 1 to I, the a^i object for the k group is represented by the assertion

$$a_k^i = \bigwedge_{j \in J_k} \left[y_j = V_{a_k^i, j} \right]$$

Formula (1) may be re-written as follows, while taking into consideration the existence of the following groups:

$$a^i = \bigwedge_{j \in J_1} \left[y_j = V_{a^i, j} \right] \wedge \cdots \wedge \bigwedge_{j \in J_k} \left[y_{j + \sum_{r<k} j_r} = V_{a_k^i, j + \sum_{r<k} j_r} \right]$$

$$\wedge \cdots \wedge \bigwedge_{j \in J_K} \left[y_{j + \sum_{r<K} j_r} = V_{a_K^i, j + \sum_{r<K} j_r} \right]$$

Objective

Like in the traditional case, the MFA based on symbolic data corresponds to the following three majors objectives:

1. Drawing up a typology of concepts, especially similar concepts from the point of view of all the variables after balancing the contribution of each group.
2. Comparing the typologies of individuals defined by groups of variables.
3. Exposing a typology of groups of variables and interpreting the proximities between them.

Table 2: Symbolic data base of boolean assertions

	y_1	y_j	y_J
C_1	$\left[x_{11}, \overline{x_{11}}\right]$	$\left[x_{1j}, \overline{x_{1J}}\right]$	$\left[x_{1J}, \overline{x_{1J}}\right]$
C_i	$\left[x_{i1}, \overline{x_{i1}}\right]$	$\left[x_{ij}, \overline{x_{iJ}}\right]$	$\left[x_{iJ}, \overline{x_{iJ}}\right]$
C_I	$\left[x_{I1}, \overline{x_{I1}}\right]$	$\left[x_{Ij}, \overline{x_{IJ}}\right]$	$\left[x_{IJ}, \overline{x_{IJ}}\right]$

However, taking into consideration the variability of objects adds an interesting element to this problem because coherence warrants us to provide exit elements in keeping with the variability.

ANALYSIS AND INTERPRETATION

We shall present algorithm for a symbolic digital MFA. This approach essentially repeats the one used by the traditional MFA as well as that of the PCA on symbolic objects. The following stages shall be used:

1) Data coding;

2) PCA of groups;

3) Weighting of the data table;

4) PCA symbolic digital of all data;

5) Traditional MFA representation and interpretation;

6) Superimposed representation of symbolic objects and average cloud;

7) Representation of assertions of variables and symbolic factors of groups

Data Coding

Either all the assertions in the SDB, the m coding order is any C application to with values in $\left(\mathbb{R}^j\right)^m$

In this work, we shall therefore use the centre and summit methods which is referred to as characteristic coding and reconstitutive coding respectively.

An assertion object is graphically represented by a hypercube. The characteristic coding represents each object by a point which is the centre of gravity of the assertion object. The reconstitutive coding represents an object by 2^j points which constitute the summit of the hypercube. Characteristic coding gives us the active elements of the analysis while reconstitutive coding gives the additional elements of the analysis that enable the reconstitution of the exit symbolic objects.

Weighting of Variables

Let X be the data table obtained after data coding. A J columns and $I \times [1+2^j]$ lines table is obtained where I individuals are active and the others in addition. For each group, a principle components analysis (PCA) on the X_k, k=1...K tables where X_k is the X table reduced to the k group value. A series of λ_i^k, i=1,...,k, k=1,...,K eigenvalues is obtained for each analysis. It is assumed that the real values in each group are classified in descending order. Each y_ℓ variable is weighted by $\dfrac{m_\ell}{\lambda_1^k}$ if the y_ℓ variable is derived from the k group. The weighted table is evaluated $X^{(p)}$.

Data Processing

Weighting conserves the internal structures of each group but changes the structure of the overall cloud (see [10]). In this case, it is referred to as the average cloud. As in the traditional case, we have represent equivalent objects by respective lines of the same K bases $\tilde{B}_k^{(p)}$ rank, with $B_k^{(p)}$ being the weighted B_k base.

$$\tilde{B}_k^{(p)} = \boxed{0 \mid B_k^{(p)} \mid 0} \Rightarrow \tilde{X}_k^{(p)} = \boxed{0 \mid X_k^{(p)} \mid 0}$$

The N_j^k K clouds may be simultaneously represented in the $\mathbb{R}^J$ space of variables by breaking down $\mathbb{R}^J$ into direct sum spaces isomorphic to $\mathbb{R}^{jk}$ spaces.

The average cloud is obtained by a homothetic transformation of the $B^{(p)}$ 1/K relation weighted data base. It represents average individuals and average concepts in $\mathbb{R}^k$. The average cloud is obtained by a homothetic transformation of the $B^{(p)}$ 1/K relation weighted data base. It represents average individuals and average concepts in $\mathbb{R}^k$. Let us note B^* the SDB of the average cloud where B_k^* constitutes its restriction in the k group.

Proposition 1. The symbolic digital PCA of the B^* base leads to the same factors as the B base characteristic coding MFA.

Proof. It is immediate. The B^* base is constructed to this effect.

Canonic Variable

The re-writing of table X^* in the principle components base $(F_1,...,F_J)$ leads to a representation of objects by a new SDB $B=\{F,O,A_F\}$ whose assertion is written

$$a_F^i = \bigwedge_{j=1}^{J} \left[F_j = W_{a_F^i, j} \right]$$

Assertions of the variables associated with concepts are written

$$F^j = \bigwedge_{i=1}^{I} \left[F_j = W_{a_F^i, j} \right]$$

These factors are called canonic variables.

Superimposed Representation of Individuals and Concepts

Equivalent Assertions

A $B^{(p)}$ based a^i assertion is represented in each $B_k^{(p)}$ sub-base by a_k^i assertions which are called equivalent assertions associated with a^i. These are the different representatives in the $B^{(p)}$ sub-bases of the same assertion.

Superimposed Representation of Individuals

Here, individuals are described by the characteristic coding. An individual representing an object is represented by the expectation of unpredictable random. variable associated with the skill it represents. This representation no doubt has some shortcomings, but it gives us the first tendencies which are sometimes enough. In concrete terms, each group is projected with additional elements on the PCA of table X^*.

Proposition 2. Each individual (average individual) of table X^* is the centre of gravity of individuals observed from the point of view of all groups.

Proof. This is the property of traditional MFA.

This proposition explains the notion of average cloud. The study of the attributes of variability of description indicated below will comfort us in this notion.

Superimposed Representation of Concepts

Symbolic digital PCA results will be used to obtain a superimposed representation of objects while taking into consideration variability. It based to the following proposition:

Proposition 3. Either $a^i = \bigwedge\limits_{j=1}^{J}\left[y_j = V_{a^i,j} \right]$ an assertion of BDS B. We note

$b^i = \bigwedge\limits_{K=1}^{K}\left[u_k = V_{b^i,k} \right]$ the assertion representing the c_i objects in the space

of the PCA principle components of this SDB. Let $G = \left(X_1^G, ..., X_J^G \right)$ the centre of gravity of the characteristic coding and u_{jk} the kth component of the jth factorial axis. Thus we have:

$$\min\left(W_{b^i,k} \right) = \sum\limits_{j,u_{kj}<0}\left[\max\left(V_{a^i,j} \right) - X_j^G \right] u_{kj}$$

$$+ \sum\limits_{j,u_{kj}<0}\left[\min\left(V_{a^i,j} \right) - X_j^G \right] u_{kj}$$

$$\max\left(W_{b^i,k} \right) = \sum\limits_{j,u_{kj}<0}\left[\min\left(V_{a^i,j} \right) - X_j^G \right] u_{kj}$$

$$+ \sum\limits_{j,u_{kj}<0}\left[\max\left(V_{a^i,j} \right) - X_j^G \right] u_{kj}$$

Proof. We note $\left(x_{ij}^G \right)_{j=1,...,J}$ the coordinate of G^i

centre of gravity of a^i assertion, and $\left(y_{kj}^G \right)_{j=1,...,J}$ the coordinate of the

image of G^i in the new basis. Let u_{jk} be the jth component of the kth

factorial axis (denoted u_k), we note $x_r^i\left(x_{ij}^r \right)_{j=1,...,J}$ the coodinate of an el-

ementary individual of a^i assertion for a point R^i, and $y_r^i\left(y_{kj}^r \right)_{j=1,...,J}$ his coordinate in the new basis.

We have:

$$y_{kj}^r = \overrightarrow{GR} \cdot u_k = \sum_{j=1}^{J} \left(x_{ij}^r - x_{ij}^G \right) u_{jk}$$

then

$$\max \left(W_{b^i.k} \right) = \max \left\{ y_{kj}^r / x_{ij}^r \in V_{a^i.j} \right\}$$

$$= \max \left\{ \left[\sum_{j=1}^{J} \left(x_{ij}^r - x_{ij}^G \right) u_{jk} \right], x_{ij}^r \in V_{a^i.j} \right\}$$

as y_{kj}^r depend of x_{ij}^r, we have

$$\max \left(W_{b^i.k} \right) = \sum_{j.u_{kj}<0} \left[\left(\min \left\{ x_{ij}^r, x_{ij}^r \in V_{a^i.j} \right\} \right) - X_j^G \right] u_{kj}$$

$$+ \sum_{j.u_{kj}>0} \left[\left(\max \left\{ x_{ij}^r, x_{ij}^r \in V_{a^i.j} \right\} \right) - X_j^G \right] u_{kj}$$

Therefore

$$\max \left(W_{b^i.k} \right) = \sum_{j.u_{kj}<0} \left[\min \left(V_{a^i.j} \right) - X_j^G \right] u_{kj}$$

$$+ \sum_{j.u_{kj}>0} \left[\max \left(V_{a^i.j} \right) - X_j^G \right] u_{kj}.$$

With the similar methode, we have

$$\min \left(W_{b^i.k} \right) = \sum_{j.u_{kj}<0} \left[\max \left(V_{a^i.j} \right) - X_j^G \right] u_{kj}$$

$$+ \sum_{j.u_{kj}>0} \left[\min \left(V_{a^i.j} \right) - X_j^G \right] u_{kj}.$$

This lead that the B_k representing group k will be projected successively with additional elements on the B^* base symbolic digital PCA. The superimposed representation on the same factorial plan enables us to generally observe the position of an object, and the point of view of each group by using variability. We then have the following result :

Proposition 4. Either a^i an assertion object represented in B^* by $(a^i)^*$.

Let us note a_k^i the a^i assertion observed from the k group point of view. In the B^* base, the a^i assertion is written

$$a^i = \bigwedge_{j \in J_1}\left[y_j = \frac{1}{J\sqrt{\lambda_1}} V_{a_1^i,j}\right]$$

$$\bigwedge \ldots \bigwedge_{j \in J_k}\left[y_{j+\sum_{r<k} J_r} = \frac{1}{J\sqrt{\lambda_k}} V_{a_k^i,j+\sum_{r<k} J_r}\right]$$

$$\bigwedge \ldots \bigwedge_{j \in J_K}\left[y_{j+\sum_{r<K} J_r} = \frac{1}{J\sqrt{\lambda_K}} V_{a_K^i,j+\sum_{r<K} J_r}\right]$$

Either $a^i = \bigwedge_{j=1}^{J}\left[y_j = V_{a^i,j}^*\right]$ we note

$a_k^i = \bigwedge_{k=1}^{K}\left[F_k = W_{a_k^i,k}\right]$ the assertion representing the c_i object from the k group point of view in the space of the PCA principle components of this SDB. Either

$G = \left(X_1^G, \ldots, X_J^G\right)$ the centre of gravity of X^*, and F_{kj} the kth component of the jth factorial axis. Thus we have:

$$\min\left(W_{a_F^i,k}\right) = \sum_{j \in J_k, F_{kj}<0}\left[\max\left(JV_{a^i,j}^*\right) - X_j^G\right]F_{kj}$$

$$+ \sum_{j \in J_k, F_{kj}>0}\left[\min\left(JV_{a^i,j}^*\right) - X_j^G\right]F_{kj}$$

$$\max\left(W_{a_F^i,k}\right) = \sum_{j \in J_k, F_{kj}<0}\left[\min\left(JV_{a^i,j}^*\right) - X_j^G\right]F_{kj}$$

$$+ \sum_{j \in J_k, F_{kj}>0}\left[\max\left(JV_{a^i,j}^*\right) - X_j^G\right]F_{kj}$$

Proof. This is the consequence of the notations of the previous proposition.

The representation of superimposed objects in the canonic base (that of factors) therefore respects the principle of centres of gravity. The hypercube representing an object is thus the centre of gravity of the hypercubes representing the same individual from the point of view of different groups. The MFA symbolic digital algorithm helps to preserve this essential property.

Representation of Variables and Assertions of Variables

The symbolic digital MFA enables us to define many representatives that contribute to the illustration of factors in the general analysis as well as in the different groups. The determination of the main factors of groups and the general analysis lead to many representations of variables and assertions of variables. The superimposed representation of these factors (as assertions of variables) on canonic variables then helps to find and interpret common factors and specific factors.

Representation of Variables and Assertions of Variables

The conduct of the MFA symbolic digital pre-supposes K+1 factorial analysis (the K groups and the average cloud). Two representations can be made for each of these analyses: a representation of variables (dual analysis of centres of gravity) and an analysis of assertions of variables taking into account variability. These representations clearly bring out the main variability factors of groups and the factors of general analysis.

Superimposed Representation of Factors (as Assertions of Variables) on Canonic Variables

This representation is the result of a MFA traditional representation which consists in highlighting the main variability factors of groups in the form of additional variables over canonic variables. But in symbolic data analysis, these variability factors are also expressed as assertions of variables and the representation of these assertions of variables over the canonic variables calls for an original observation. The position and collection of assertions of variables give us precious indications on their importance and the link between the different groups.

Link between Groups: Representation Quality of Groups

This is a traditional indicator in MFA. It is a representation of groups on canonic factors. The coordinates of an axis group is the combined inertia of the group's variables on the corresponding MFA axis of table X^*. For the two groups K_x. and K_y, we have

$$\mathcal{L}\left(K_x, K_y\right) = \left\langle W_x^{(p)} D, W_y^{(p)} D \right\rangle$$

where D is the diagonal

$I \times I$ matrix of the weight of individuals and $W_j^{(p)}$ is the matrix of scalar products between the individuals of group j for the X_j^p weighted matrix.

A symbolic representation may be proposed for it. Each assertion of variables can be coded into many variables whose projection over canonic variables results in a minimum value and a maximum value. The variability associated to a group will therefore be displayed by two values: the sum of the minimum inertia of the group's variables and the sum of its maximum inertia.

The representation of groups is done in the $\mathbb{R}^{I^2}$ space.

The Notion of Common Factor and Specific Factor

MFA helps to portray the characteristics of groups compared to the canonic variables (also referred to as factors). A common (or specific) factor to the MFA symbolic digital will be the common factor associated to the traditional MFA of centres of gravity.

POVERTY DATA

Presentation

The data is drawn from the first Cameroonian survey in households (ECAM I) conducted in 1996 by the National Institute of Statistics. The sample is made up of 1800 households with a total of 10230 individuals. The analysis is done on 13 continuous variables out of which 10 are active and 3 illustrative (see Table 3).

The average poverty line is used to establish the difference between poor and non poor individuals. We shall consider 10 classes of individuals with 5 poor classes organized in descending order from very poor to less poor individuals and 5 classes of non poor individuals organized in ascending order from less rich to very rich individuals. These classes constitute the concepts that we shall analyze. The groups of variables are regions. For presentation purposes, they shall be presented on a line and not in columns. An overview of the SDB for the Adamawa Region for three variables shows the following Table 4.

Analysis and Interpretation

After the algorithm application, the following results are obtained:

Inertia of the First Six Eigenvalues

Please see Table 5.

Relationship between General Variables and Groups

The F_1 variable is the variable which extracts the most important inertia from all groups. The other variables convey specific information to certain regions (see Table 6).

Table 3: Variables of ECAM1

N	Title	Description	Quality
1	DEPEAD	Expenditure per adult equivalent	Additional
2	DEPTET	Expenditure per head	Additional
3	DEPTOT	Total expenditure	Additional
4	DEPALI	Food expenditure	Active
5	DEPHAD	Clothing expenditure	Active
6	DEPMAI	Household expenditure	Active
7	DEPSAN	Health expenditure	Active
8	DEPTRA	Transportation expenditure	Active
9	DEPEDU	Education expenditure	Active
10	DEPSOP	Personal health care expenditure	Active
11	DEPLOI	Leisure expenditure	Active
12	DEPLOY	Rent expenditure	Active
13	DEPLOG	Lodging expenditure	Active

Table 4: An overview of SDB of ECAM1

Concept	DEPALI	AD		
		DEPHAB	DEPMAI	
Poor of class 1	[96,464, 173,375]	[2000, 36,000]	[3950, 36,000]
Poor of class 2	[122.375, 573,832]	[15,000, 31,200]	[28,800, 34,800]
Poor of class 3	[109,500, 573,032]	[5700, 46,800]	[10,800, 28,800]
Poor of class 4	[157,210, 302,428]	[5700, 38,200]	[7200, 15,300]
Poor of class 5	[126,967, 358,482]	[0, 54,500]	[12,000, 53,700]
Non poor of class 1	[102,982, 536,028]	[0, 115,900]	[2450, 55,100]
Non poor of class 2	[220,303, 568,617]	[0, 82,000]	[11,950, 57,600]
Non poor of class 3	[95,421, 1,707,678]	[0, 297,300]	[23,000, 65,100]
Non poor of class 4	[146,782, 967,510]	[3500, 297,300]	[600, 71,300]
Non poor of class 5	[130,617, 813,428]	[8000, 485,200]	[22,800,199,800]

Table 5: Histogram of eigenvalues

	Eigenvalue	Inertia %	Combined %
1	6.80938	68.09	68.09
2	1.08859	10.89	78.98
3	0.60479	6.05	85.03
4	0.56497	5.65	90.68
5	0.31109	3.11	93.79
6	0.29054	2.91	96.69

Table 6: Relationship between general variables and groups

	F_1	F_2	F_3	F_4	F_5	F_6
AD	52.70	20.95	5.81	3.94	1.15	1.31
CE	77.57	9.29	0.83	0.93	0.35	0.26
ES	58.61	4.60	9.72	8.58	0.96	4.04
EN	62.35	10.50	3.13	5.98	2.81	1.40
LT	82.74	2.18	1.31	1.93	1.09	0.45
NO	51.03	6.12	7.38	8.56	10.40	4.10
NW	67.16	8.38	2.91	5.74	1.36	3.81
OU	46.08	10.14	6.78	9.28	6.78	5.22
SU	57.94	13.26	523	2.32	2.02	3.20
SW	57.84	12.42	11.17	3.22	1.13	2.62

Inter-Inertial/Total Inertia Relationship

Table 7 shows that on the first axis, there is a close proximity between equivalent classes. Broadly speaking, these equivalent classes have similar behaviours in their expenditure habits. The same phenomenon is observed in the descending values of r, on the 4th, 2nd, 6th and 3rd axes.

Degree of Resemblance between Groups and General Variables: Common and Specific Factors

The first MFA component is a factor common to all the 10 regions. The second factor is common to the FarNorth and South-West regions. The 3rd factor is specific to the South-West and Adamawa regions (see Table 8).

Table 7: Inter-inertial/total inertia relationship

Axis	F_1	F_2	F_3	F_4	F_5	F_6
Relationship	0.89	0.70	0.64	0.70	0.30	0.65

Table 8: Degree of resemblance between groups and general variables

	F_1	F_2	F_3	F_4	F_5	F_6
AD	0.87	0.58	0.65	0.13	0.05	0.09
CE	0.90	0.12	0.03	0.45	0.01	0.02
ES	0.90	0.15	0.31	0.66	0.15	0.80
EN	0.98	0.75	0.45	0.50	0.18	0.38
LT	0.98	0.28	0.11	0.33	0.02	0.08
NO	0.92	0.42	0.44	0.32	0.59	0.32
NW	0.97	0.57	0.12	0.78	0.28	0.35
OU	0.88	0.42	0.43	0.60	0.28	0.62
SU	0.93	0.47	0.60	0.55	0.28	0.70
SW	0.96	0.68	0.72	0.30	0.15	0.16

REFERENCES

1. E. Diday, "Introduction to the Symbolic Data Analysis," Cahiers du CEREMADE, No. 8823, 1988.
2. E. Diday, "From the Objects of the Data Analysis to Those of the Analysis of Knowledge, Symbolic and Numeric Induction from Data," Cépadues, 1991.
3. L. Lebart, A. Morineau and Mr. Piron, "Multidimensional Exploratory Statistics," 3rd Edition, DUNOD, Paris, 2000.
4. M. Volle, "Data Analysis," Economica, Paris, 1981.
5. J. F. Martin, "Fuzzy Coding and Its Applications in Statistics," Thesis, University of Pau, Pau, 1980.
6. F. J. Gallego, "Fuzzy Coding in Correspondence Analysis," Cahiers de l'Analyse des Données, Vol. 13, No. 2, 1980, pp. 413-430.
7. J. Pagès and B. Escofier, "Simple and Multiple Factorial Analysis," DUNOD, Paris, 1990.
8. B. Escofier and J. Pajes, "Multiple Factorial Analysis, Objectives, Methodology and Interpretation," 2nd Edition, Wiley, Paris, 2005.
9. P. Cazes, A. Chouakria, E. Diday and Y. Schektman, "Extension of Principal Component Analysis to Interval Data," Revue de Statistique Appliquée, Vol. 45, No. 3, 1997, pp. 5-24.
10. B. Ahanda Tang, "Extension of Factorial Analysis to Symbolic Data," Thesis, University of Paris, Paris, 1998.

CITATION

B. Ahanda, J. Jiofack, R. Nzangué and G. Mbiakop, "Multiple Factorial Analysis of Symbolic Data," Applied Mathematics, Vol. 3 No. 12A, 2012, pp. 2148-2154. doi: 10.4236/am.2012.312A295.

The P≠NP Conjecture in the Context of Real and Complex Analysis

Jerzy Mycka[a] and José Félix Costa[b]

[a]Institute of Mathematics, University of Maria
Curie-Sklodowska, Lublin, Poland
[b]Department of Mathematics, I.S.T.,
Universidade Técnica de Lisboa, Lisboa,
Portugal

ABSTRACT

In this paper, we aim at an analog characterization of the classical P
≠ NP conjecture of Structural Complexity. We consider functions over
continuous real and complex valued variables. Subclasses of functions
can be defined using Laplace transforms adapted to continuous-time
computation, introducing analog classes DAnalog and NAnalog. We
then show that if DAnalog ≠ NAnalog then P ≠ NP.

INTRODUCTION AND MOTIVATION

We have been working towards recursive definitions of computational
classes of functions over . The first presentation of such a theory, analogous to Kleene's classical theory of recursive functions over , was at-
tempted by Cristopher Moore [9]. Real recursive functions are generated by a fundamental operator, called differential recursion. The other
fundamental operator is the taking of infinite limits, introduced in [10].
In [5] one of the authors, along with Campagnolo and Moore show
that a linear form of the differential recursion scheme gives rise to an
analog characterization of the class of (Kalmar's) elementary functions
and to a general analog characterization of the Grzegorczyk hierarchy.
In [4,6] it is shown that the GPAC is not closed under iteration and that
a subclass of real recursive functions coincides with the class of GPAC-
computable functions. In [11], we finally show how to capture higher

computational classes through limit operators. Manuel Campagnolo showed also in [3] that other computational complexity classes can be captured through appropriate structured differential schemata or by adding simple or bounded integration. Recently, Olivier Bournez and Emmanuel Hainry proved in [2] that a specific kind of limit operator together with differential recursion makes the class of real recursive functions an exact extension to the real numbers (we can say in the sense of Computable Analysis) of the classical recursive functions.

We believe that the theory of real recursive functions, with infinite limits [11], has enough ingredients to allow a good translation of classical computability and classical computational complexity problems into Analysis. We do believe that such translations might be a solution to open problems described in analytic terms: in this paper, we are much involved in the definition of some analog classes, and to find one analytic representation of the conjecture $P \neq NP$. The main goal will be to connect this open problem with problems of Analysis.

We introduce in this paper the classes DAnalog and NAnalog, which are subclasses of real recursive functions defined using the Laplace transform. Let us point out that their names do not indicate time or space complexity, however we will connect DAnalog with PF and NAnalog with NPF (classes PF and NPF will be introduced shortly: they stand for functions computable in polynomial time by deterministic and non-deterministic Turing machines, respectively). Let us add that it is not our purpose to build such classes, which strictly inherit the properties of the classical complexity classes; hence, for example, the use of limits in DAnalog is not forbidden for us. We want to find the classes, which are interesting from a point of view of analog computation and such that some relations between these classes in the analog world are analogous to relations between their counterparts in the discrete case.

GROWTH OF FUNCTIONS AND SUBEXPONENTIALITY

We start by recalling that the computational complexity of functions over the non-negative integers is connected with their rate of growth. Let us look at this definition (see [15]).

Definition 1: A function h of arity n is defined from a function f of arity n and a function g of arity (n + 2), by polynomially bounded primitive recursion, if there are polynomials p and q, and a function t of arity (n + 1), such that [1]

$$h(x_1, \ldots, x_n) = t\left(x_1, \ldots, x_n, p\left(\sum_{i=1}^{n} \lfloor \log(x_i + 1) \rfloor\right)\right),$$

where

$$t(x_1, \ldots, x_n, 0) = f(x_1, \ldots, x_n),$$

$$t(x_1, \ldots, x_n, y + 1) = g(x_1, \ldots, x_n, y, t(x_1, \ldots, x_n, y)),$$

such that the following condition holds: for all

$$y \leqslant p(\sum_{i=1}^{n} \lfloor \log(x_i + 1) \rfloor),$$

we have

$$\lfloor \log(t(x_1, \ldots \ldots x_n, y)) \rfloor \leq q(\sum_{i=1}^{n} \lfloor \log(x_i + 1) \rfloor)^2$$

The function log can here be used to measure the size of the inputs represented by bit sequences: if x is one of the inputs, then it can be expressed in $0 \lfloor \log(x + 1) \rfloor$ bits. In this section we could have used, as usual, $|x|$ for the size of x. Since the size of x is approximately given by $\lfloor \log(x + 1) \rfloor$ we will use this intuition as a guideline for next sections, while working with real-valued functions. The size of all inputs taken together is $o(\sum_{i=1}^{n} \lfloor \log(x_i + 1) \rfloor)$

Let us recall that the main model of computation, the Turing machine, has an obvious correspondence with the class of recursive functions. Hence, we can distinguish within the set of recursive functions a subset PF corresponding to deterministic Turing machines working in polynomial time.

A partial function f is said to be computable by a deterministic Turing machine M if (a) M accepts the domain of f and (b) if $\langle x_1,\ldots\ldots\ldots x_n \rangle \in$ dom(f), then the accepting computation writes in the output tape the value $f(x_1,\ldots,x_n)$.

Definition 2. PF is the class of partial functions that can be computed in polynomial time by deterministic Turing machines, i.e., by deterministic Turing machines clocked with polynomials.

We adopted the definition of partial recursive function given in [1]. Note that for functions in PF we can always check the domain, hence the halting problem in this case is decidable. This problem arises because we use partial functions; partial functions are simpler to handle the no determinism. We have an inductive definition of the total functions in PF, provided by Buss in 1986 (see, e.g., [15, p. 172]):

Proposition 3. The class of total recursive functions computable in deterministic polynomial time is inductively defined from the basic functions $Z = {}_n.0$ and $S = {}_{n.n+1}$, the projections, basic functions ${}_n. 2n, {}_n. 2n + 1, n. [n/2]$, the characteristic function of 'equality to 0', by the operations of composition, definition by cases, and polynomially bounded primitive recursion.

The intuition behind this characterization of the total functions in PF is the following: a Turing machine clocked in polynomial time p can write at most $p(|x|)$ bits in the output tape. This number is bounded by $2^{|x|k}$ for some k.

For our purposes, we define a non-deterministic Turing machine in a way similar to which it is used in the probabilistic computational model. We use the notion of a time constructible function: a function is time-constructible if there exists a Turing machine such that for each input x of length n it halts in an accepting state after f(n) steps of computation. We impose the following conditions on non-deterministic machines (see [1] for the probabilistic Turing machine): (a) every step of a computation can be made in exactly two possible ways, which are considered different even if there is no difference in the corresponding actions (this distinction corresponds to two different bit guesses), (b) the machine is clocked by some time-constructible function and the number of steps in each computation is exactly the number of steps allowed by the clock; if a final state is reached before this number of steps, then the computation is continued, doing nothing up to this number of steps, (c) every computation ends in a final state, which can be either accept or reject, (d) if the machine computes a function, then all accepting computations write down to the output tape the value of the function (see [1, Chapter 2], for a comparison). It is irrelevant what the machine writes in the output tape if it reaches a rejecting state.

A partial function f is said to be computable by a non-deterministic Turing machine M if (a) M accepts the domain of f and (b) if $\langle x_1, \ldots \ldots x_n \rangle$ ∈ dom (f), then any accepting computation writes on the output tape the value f $(x_1, \ldots, x_n)$.

Definition 4: NPF is the class of partial functions that can be computed in polynomial time by non-deterministic Turing machines (i.e., by non-deterministic Turing machines clocked with polynomials).

Definition 5: We define the class ∃PF as follows: for a function f : ^{n} → in ∃PF there exists a function F : $^{n+1}$ → in PF and a polynomial p : ^{n} → such that (a) $\langle x_1, \ldots \ldots x_n \rangle$ ∈ dom() if and only if there exists a number k such that $|k| \leq p(|x_1|, \ldots, |xn|)$ and $\langle x_1, \ldots \ldots x_n, k \rangle$ ∈ dom(F) and (b) f $(x_1, \ldots, x_n)$ is defined and f $(x_1, \ldots, x_n) = y$ if and only if there exists a number k such that $|k| \leq p(|x_1|, \ldots, |x_n|)$, F $(x_1, \ldots, x_n, k)$ is defined and F $(x_1, \ldots, x_n, k) = y$, and, for all such $|k| \leq p(|x_1|, \ldots, |x_n|)$ such that $\langle x_1, \ldots \ldots x_n, k \rangle$ ∈ dom(F), we have F $(x_1, \ldots, x_n, k) = y$.

With the above definitions the following fact can be easily proved.

Proposition 6: The class NPF coincides with ∃PF.

From the last proposition we also have that the function f has domain in NP if and only if the function F has domain in P, fulfilling the classical theorem NP = ∃P: a set A is in NP if and only if there exists a polynomial p and a set B in P such that x ∈ A if and only if ∃$|z| \leq p(|z|)$ ($\langle x, z \rangle$∈ B).

A characteristic function of a set A will be the partial function c_A defined by the expression: $c_A(x) = 1$ if x ∈ A, otherwise it is ⊥. For sets in P this is like defining a total function since the domain of a characteristic function is always decidable (the value ⊥ is then interpreted as 0).

The following statements, although they are about classical computational complexity, they are (to our knowledge) not found elsewhere for the definitions as given here. Let us start with an obvious result.

Proposition 7: A set is in P if and only if its characteristic function is in PF. A set is in NP if and only if its characteristic function is in NPF.

We can observe that, of course PF $\subseteq$ NPF. Now let us connect classes of functions and sets.

Proposition 8: NP $\subseteq$ P if and only if NPF $\subseteq$ PF

The above proposition can be proved by a simple construction, as a consequence of the two main propositions in this section we can state immediately:

Proposition 9: P = NP if and only if PF = NPF if and only if NPF = $\exists$PF.

We know now that whenever the conjecture P $\neq$ NP is in context, then it can be replaced by the problem PF $\neq$ NPF. In our framework of computation over, we deal with functions rather then sets. Hence, we translate a problem about sets (classes of sets) into a problem about functions (classes of functions).

Before considering the next concepts, we should strongly stress that functions in PF or in NPF can be considered total. The reason is obvious: whenever a function f is undefined at x we can give to f (x) value 0 (remember that the domain of f is decidable). However, in the particular case of a function f in PF, f can always be considered algorithmically total in PF

$$new[f](x) = \begin{cases} f(x) + 1 & \text{if } x \in \text{dom}(f), \\ 0 & \text{otherwise.} \end{cases}$$

Since for f in PF, dom(f) is in P, the function new[f](x) is in PF.

It is also interesting to observe that NPF is the class of functions of the form $x_1 \ldots x_n$.if $\langle x_1, \ldots \ldots x_n \rangle \in A$ then F $(x_1 \ldots x_n)$, where A $\in$ NP and F $\in$ PF.

We end this section with the following definition:

Definition 10: A quasi-polynomial (sometimes called weak exponential) is a function from the class of functions

$$\{2^{\log(x+1)^k} : k \in \mathbb{N}\}.$$

RECURSIVE FUNCTIONS OVER AND OVER

We have introduced in [11] the concept of a real recursive function (modifying the original definition of R-recursive functions in [9]) and the corresponding class REC($\mathbb{C}$); here we will define the concept of complex recursive function and the corresponding class REC($\mathbb{C}$), needed to work with the Laplace transform.

Definition 11: REC ($\mathbb{C}$) The class REC ($\mathbb{C}$) of complex recursive vector functions is generated from the complex recursive scalars 0, 1, i, and the complex recursive projections I_n^j ($z_1,..., z_n$) = z_j, $1 \le j \le n$, $n > 0$, by the following operators:

Composition: if f is a complex recursive vector function with n k-ary components and g is a complex recursive vector function with k m-ary components, then the complex vector function with n m-ary components, $1 \le j \le n$,

$$\lambda x_1 \ldots \lambda x_m . \; f_j(g_1(z_1, \ldots, z_m), \ldots, g_k(z_1, \ldots, z_m))$$

is complex recursive.

Differential recursion: if f is a complex recursive vector function with n k-ary components and g is a complex recursive vector function with n (k + n + 1)-ary components, then the complex vector function h of n (k + 1)-ary components which is the solution of the Cauchy problem, $1 \le j \le n$,

$$h_j(z_1, \ldots, z_k, 0) = f_j(z_1, \ldots, z_k),$$

$$\partial_z h_j(z_1, \ldots, z_k, z) = g_j(z_1, \ldots, z_k, z, h_1(z_1, \ldots, z_k, z), \ldots, h_n(z_1, \ldots, z_k, z))$$

is complex recursive if, for all scalar components, their real and imaginary parts are of the class C^1 on the largest interval containing $(0, 0)$ in which a unique solution exists.

Infinite limits: if f is a complex recursive vector function with n (k + 1)-ary components, then the complex vector functions h, h^{inf}, h^{sup} with n k-ary components, $1 \leq j \leq n$,

$$h_j(z_1, \ldots, z_k) = \lim_{z \to \infty + i0} f_j(z_1, \ldots, z_k, z),$$

$$h_j^{inf}(z_1, \ldots, z_k) = \liminf_{z \to \infty + i0} f_j(z_1, \ldots, z_k, z),$$

and finally

$$h_j^{sup}(z_1, \ldots, z_k) = \limsup_{z \to \infty + i0} f_j(z_1, \ldots, z_k, z)$$

are complex recursive vector functions.

Assembling and Designating Components: (a) arbitrary complex recursive vector functions can be defined by assembling scalar complex recursive scalar components; (b) if f is a complex recursive vector function, then each of its scalar components is a complex recursive scalar function

Complex recursive numbers: arbitrary complex recursive scalar functions of arity 0 are called complex recursive numbers.

In the above definition, we can take limits (either lim, or lim inf, or lim sup) to $\infty + i0$, but a simple change of variables will allow us to take limits to $x_0 + i\infty$, where x_0 is a real recursive number. Limits to $x_0 + i\infty$ are particularly important to compute, e.g., the inverse Laplace transform, Euler's é function, Riemann's ζ function, etc.

The restriction of the domain and range in the above definition to leads us to the concept of a real recursive function, which is presented in detail in [11]. Here we use only slightly informal definition.

Definition 12: The class REC() of real recursive vector functions is defined as in Definition 11 with all domains and ranges restricted to and with the consequent obvious modifications.

Let us only recall some useful results. First, the functions x +y, x ×y, x −y, e^x , sin x, cos x, $\frac{1}{x}$, x/y, log x, x^y are real recursive functions. Also some special functions are real recursive too: the Kronecker function, the signum function, and absolute value, the Heaviside function (equal to 1 if x 0, otherwise 0), the binary maximum max, the square-wave function s (s(x) = 1 for x ∈ [2_n, 2_n + 1], n ∈ , and s(x) = 0 otherwise), and the floor function. Particularly interesting real recursive functions from REC() are functions obtained by the iteration functional. We give the following proposition after [9,11] (where the proof can be found).

Proposition 13: If f is a real recursive function of arity one, then the iteration of f, F, is a real recursive function of arity two, such that, for all n ∈ , F (n, x) = $f^n(x)$ (i.e., the iteration functional preserves real recursiveness).

The last result provides a binary real recursive function F such that, on non-negative integers in the second argument, F gives the iteration function $_{xn}$. $f^n(x)$. If x is a binary representation of some number and p is a polynomial on |x|, then F (x, p(|x|)) represents the value after a polynomial running of f

Proposition 14: If f : ^{k} → is a partial recursive function, then there exists a real valued function F : ^{k} → such that

1. F is a real recursive,

2. for all

$x_1, \ldots, x_k \in \mathbb{N},$

We have

$F(x_1, \ldots, x_k) = f(x_1, \ldots, x_k),$

3. for all

$x_i \in [n_i, n_i + 1], n_i \in \mathbb{N}, 1 \leqslant i \leqslant k,$

We have

$$\min(F(n_1, \ldots, n_k), \; F(n_1 + 1, \ldots, n_k + 1)) \leqslant F(x_1, \ldots, x_k)$$
$$\leqslant \max(F(n_1, \ldots, n_k), \; F(n_1 + 1, \ldots, n_k + 1)).$$

Proof: Let us start with the simple observation that the points 1 and 2 of the proposition can be obtained via the simulation of Turing machines by real recursive functions (e.g., such as in [9, 11]).

Let F': $_k \rightarrow$ be such a function. The third point can be established by a simple transformation: $F(x_1, \ldots, x_k) = F' (\lfloor x_1 \rfloor, \ldots, \lfloor x_k \rfloor)$.

For a given function $f : N^k \rightarrow N$ we will call its counterpart described in the above manner a canonical extension.

We recall from classical recursion theory that the recursion scheme can be substituted by iteration if we join to the set of primitive functions the functions to code and decode pairs of numbers (see [14]). We can add the following theorem, which is based on [4].

Proposition 15: Let $f : ^k \rightarrow$ be a primitive recursive function. Then its canonical extension $F : ^k \rightarrow$ can be defined by composition and differential recursion from the basic functions -1, 0, 1, Θ, and the projections.

In [2] is introduced a minimalization operator in order to capture, together with composition and differential recursion, exactly the whole class of partial recursive functions and no more integer valued functions. For the same reasons, expressed in the paragraph above, here we prefer the concept of limit as being efficient and allowing us to reason about computability and complexity in Analysis in a natural way.

Now let us consider a complex domain. Here the functions $z_1 + z_2$, $z_1 \times z_2$, $z_1 - z_2$, e^z, $\sin z$, $\cos z$, $\sinh z$, $\cosh z$, $\frac{1}{z}$, z_1/z_2, $\log z$, z^w, $\tan^{-1}(z)$ are complex recursive. We present here only a few justifications. Trigonometric functions can be defined by composition, using the formulas

$$\sin(z) = \frac{e^{iz} - e^{-iz}}{2i} \text{ and } \cos(z) = \frac{e^{iz} + e^{-iz}}{2}.$$ In the case of the hyperbolic

functions, we use the formulas $\sin(z) = \dfrac{e^z - e^{-z}}{2}$ and $\cos(z) = \dfrac{e^z + e^{-z}}{2}$.

The functions sin and cos can be also obtained directly as follows: $\sin(z) = -i \sinh(iz)$ and $\cos(z) = \cosh(iz)$. The rest of definitions can be done in the standard way (see [11]).

By structural induction we can also prove the following theorems (the second one implies that REC() ⊆ REC() in some sense)

Proposition 16: If a function $f : \rightarrow$ is a complex recursive function, f $(z) = f(x, y) + i f_{\Im} (x, y)$, where $z = x$ and $\Im z = y$, then f and f are real recursive functions.

Proposition 17: If $f : \rightarrow$ is a real recursive function in REC(), then there exists a function $f^ : \rightarrow$ in REC() such that its imaginary part at $x + i0$ vanishes and the following equation holds

$$f(x) = \hat{f}(x + i0).$$

Let us add a digression about using the Laplace transform and its inverse in the context of real and complex recursive functions. The formulas

$$F(s) = \int_0^\infty f(\xi)\, e^{-s\xi}\, d\xi, \quad f(\xi) = \frac{1}{2\pi i} \int_{c-i\infty}^{c+i\infty} F(s)\, e^{s\xi}\, ds$$

are referred to as the Laplace transform $F = L[f]$ and the inverse Laplace transform $f = L^{-1}[F]$, respectively. The second integral is generally carried out by a contour integration. Following [16], consider a real recursive function f on the positive real axis such that: (i) f is continuous on $[0, \infty)$ except possibly for a finite number of jump discontinuities in every finite sub-interval; (ii) there is a positive number M such that the absolute value $|f(\xi)| \leq M e^{k\xi}$, for all ≥ 0, in which case we say that f belongs to the class L_k. Additionally let $L = \bigcup_{k>0} L_k$. Then we can cite the following after [12].

Proposition 18: If $f \in L_k$ is a real recursive function, then the Laplace transform $L[f](x + iy)$ exists for $x > k$ and it is complex recursive.

From the following infinite limits of functions defined by differential recursion (but given by simple integration)

$$\lim_{\tau \to \infty} \int_0^\tau e^{-s\xi}\, f(\xi)\, d\xi$$

and

$$\lim_{\tau \to 0 + i\infty} \int_{c-\tau}^{c+\tau} e^{\xi z} \, F(z) \, dz$$

we can obtain the following result.

Proposition 19: The Laplace transform and the inverse Laplace transform are, respectively, real recursive and complex recursive functions, whenever their function arguments are real recursive and complex recursive functions, respectively.

We are now going to show, how to use the Laplace transform together with the Bromwich contour to get a good understanding of the class of recursive functions over the reals and over the complexes. We also show that almost all real recursive functions obtained and used to our purpose in previous papers (e.g., [9,5,6,11,12]) can be obtained in a very elegant and ingenious way.

The guideline is as follows: taking a real recursive function f (x) we obtain its complex recursive transform F (s), from which we obtain a real recursive function F (x + i0), and then we apply again the machinery or combining two complex recursive functions F1(s) and F2(s), we find a new complex recursive function F (s), from which we obtain a real recursive function using the Bromwich contour. In order to use the Bromwich contour, we have to ensure that the given function F of the complex variable s is analytic throughout the finite s plane except for a finite number of isolated singularities.

We start with the Heaviside step function Θ. The Laplace transform is the complex recursive function $L[\Theta(x)](s) = \dfrac{1}{s}$. From this function we conclude that also $_x \cdot \dfrac{1}{x}$ is also real recursive. Since simple integration of a real recursive function is a real recursive function, integrating (x) we obtain two new real and one new complex recursive functions x

$$x = \int_0^x \Theta(\xi)\,d\xi, \quad L\left[\int_0^x \Theta(\xi)\,d\xi\right](s) = \frac{L[\Theta(x)](s)}{s} = \frac{1}{s^2}.$$

Iterating this procedure we get more recursive functions: starting with (x) we reach two new real and one new complex recursive functions:

x^n, $\frac{1}{x^{n+1}}$. From this reasoning it also follows that the function $_x\cdot\frac{1}{x^n}$ is real recursive, for all $n > 0$.

In general, when we have a periodic function f such that $f(x) = f(x+a)$, its Laplace transform is given by

$$\frac{\int_0^a e^{-s\xi} f(\xi)\,d\xi}{1 - e^{-as}}.$$

Using the Bromwich contour to transform back the complex recursive

function $_s\,\frac{1}{s^2} \times \tanh s$ we get the saw-tooth function

$$st(x) = x\Theta(x) + 2\sum_{n=1}^{\infty} (-1)^n\,(x - 2n)\,\Theta(x - 2n).$$

The square-wave and the saw-tooth functions are used in [4,9,11] to obtain the iteration of a real recursive function as real recursive function. From the iteration of a real recursive function we can then get the simulation of the execution of a Turing machine.

As a final remark, we note that many periodic discontinuous functions can be obtained using the method just described. However, we should keep in our minds the following statement. If f and g both belong to L, and F (s) = G(s) for all (sufficiently) large values of s, then f (t) = g(t) for all values of t where f and g are continuous. The last statement (taken from [16]) shows that the above mentioned functions are defined up to a countable number of discontinuity points, a set of measure zero, that does not interfere with the solutions of differential equations involving such functions.

The consideration of Turing machines in the context of real or complex recursive functions is not of utmost importance, since we already have a characterization of the classes of functions computed in polynomial time by deterministic and non-deterministic Turing machines using recursive function theory. However, we would like to justify our definition in this aspect too.

Let us recall that the set of possible computations of a non-deterministic Turing machine M on an input string can be described by a binary tree of computations: the nodes of the tree are configurations of the machine M on input; the root is the initial configuration, and at any node, its sons are those configurations which can be reached in one move. In this way, we can simulate a run of a non-deterministic Turing machine M, giving to M an input word x together with a guess word y, i.e., two real numbers, which code for the input and the guess, respectively. Note that, given a guess y, then after retrieval of the last bit

$b := b(y)$, the remaining part is given by $r := \dfrac{y - b(y)}{2}$. This function is analytic if we find an analytic way of determining $b(y)$. We know that, as consequence of our choice of binary alphabet, the last bit of the guess is just

$$b := 1 - \frac{\cos(\pi y) \times (1 + \cos(\pi y))}{2}.$$

We can then look at b as a function of real y. This real function for integer arguments gives b, which is a bit.

Using the analytic map of two dimensions introduced in [8] and extending it with the guess we find that there are analytic real recursive functions which can simulate the transition of any non-deterministic Turing machine. All what is needed to justify this statement is the additional component of f_M to use a variable for non-determinism:

$$[f_M(x_1, x_2, y)]_3 := \frac{y - b(y)}{2}.$$

To obtain the function computed by M it is enough to iterate the steps until we reach the final state. We can use limits here: if the final state is found then the sequence $\left\{ \int_M^n (x_1, x_2, y) \right\} n \in \mathbb{N}$ has a limit, otherwise it diverges. Let us add that the halting problem for Turing machines is decidable by a real and a complex recursive function (see [9,11]). However, it does not mean that all problems are decidable by means of real recursive functions. The full discussion and examples of the problems undecidable by means of real (and complex) recursive functions is given in [13].

DANALOG AND NANALOG

A real (or complex) total recursive function is said to be of exponential order if in every step of its construction, its components (the components of their real and imaginary parts in the complex case) are of exponential order. It is said to be of subexponential order if in every step of its construction, its components (the components of their real and imaginary parts in the complex case) are of subexponential order

In Analysis we have a precise boundary between two different worlds of computable functions. The subexponential world has the following property: for every function f, $L[f](s)$ is defined along the whole positive real axis $s > 0$. The exponential world has the weaker property: for every function f, $L[f](s)$ is defined for values of the complex variable s such that $s \geq x_f$, where x_f depends on f.

To consider some functions as subexponential, sometimes we have to make a small shift on the real variable to avoid a discontinuity at the origin. E.g., function $_x.\dfrac{1}{x+\varepsilon}$ is subexponential and its transform is $_s.$ $e^{\varepsilon s} E_1(_s)$, where E_1 is the exponential integral of degree one, for positive as small as we want.

Trivially, it can be shown that:

Proposition 20: Subexponential functions are preserved by differentiation

Proof. We will prove that if f is a total function of x, such that its Laplace transform is defined for the all positive real axis, then the function f¯ defined by $f^-(x) = \partial_x f(x)$ is subexponential too. We have $f(x)L[\lambda_x \partial_x f(x)](s) = sF(s) - f(0)$ whenever $L[f] = F$. This fact completes our proof, since $\lambda_s. s F(s)$ is defined for all the positive real axis, whenever F is defined too in the same open interval

Proposition 21: Subexponential functions are preserved by integrals

Proof. We will prove that if f is a total function of x, such that its La-place transform is defined for the all positive real axis, then the function f⁻ defined by $f(x)^- = \int_0^x f(\xi)\,d\xi$ is subexponential too. Let f be a real valued function of ξ. We have $L[\int_0^x f(\xi)\,d\xi](s) = \frac{F(s)}{s}$ whenever $\frac{F(s)}{s}$. This fact completes our proof, since F (s) s is defined for all the positive real axis, whenever F is defined too in the same open interval.

Proposition 22: Subexponential functions are preserved by limits (i.e. if the limit of a subexponential function exists, then it is subexponential).

Proof: We have to prove that if f is a total function of x and y, such that their Laplace transforms with respect to x and y are defined for all posi-tive reals, then the function f⁻ defined by $f(x)^- = \lim_{\to\infty} f(x,)$, if such a limit exists, is subexponential too. Consider f, a real valued function of x and υ, and any infinite divergent sequence υ_n of real numbers; we have, for all such sequences υ_n, assuming that $f_n(x) = f(x, \upsilon_n)$ is mea-surable, [3]

$$L_x\left[\lim_{n\to\infty} f_n(x)\right](s) = \int_0^\infty \lim_{n\to\infty} e^{-s\xi} f(\xi, \upsilon_n)\,d\xi = \lim_{n\to\infty} \int_0^\infty e^{-s\xi} f(\xi, \upsilon_n)\,d\xi$$
$$= \lim_{n\to\infty} L_x[f_n(x)](s).$$

This means that $L_x[\lim_{\upsilon\to\infty} f(x,\upsilon)](s) = \lim_{\upsilon\to\infty} L_x[f(x,\upsilon)](s)$, i.e., $L_x[\lim_{\upsilon\to\infty} f(x,\upsilon)](s) = \lim_{\upsilon\to\infty} L_x[f(x,\upsilon)](s) = \lim_{\upsilon\to\infty} F_x(s,\upsilon)$ with $L_x[f(x,\upsilon)](s) = F_x(s,\upsilon)$. using the Bromwich contour we obtain, in similar lines,

$$\frac{1}{2\pi i} \int_{c-i\infty}^{c+i\infty} \lim_{\upsilon\to\infty} F_x(s,\upsilon)\, e^{s\xi}\,ds = \lim_{\upsilon\to\infty} f(x,\upsilon)$$

that is $\lim_{\upsilon \to \infty} f(x, \upsilon) = L^{-1}[\lim_{\upsilon \to \infty} F_x(s, \upsilon)](x)$, and, since the limit in the left hand expression exists and F_x is defined, the limit in the right hand must also exist, meaning that, $\lim_{\upsilon \to \infty} f(x, \upsilon)$ is subexponential too.

We propose the definition of the classes DAnalog and NAnalog, which can be interpreted as classes of real recursive functions computed with quasi-polynomial restrictions on their values. The definitions are not suggested by the classical counterparts, but they will be strongly motivated. Of course, according to what was said above, the classes DAnalog and NAnalog will be defined as subclasses of REC($\mathbb{R}$).

Definition 23. The class of real recursive functions, designated by DAnalog, is defined as the collection of all real recursive functions f such that:

1. $L[f]$ is defined on the whole positive axis;
2. f is 0, 1, −1, θ_2 ($\theta_2(x) = x^2 \Theta(x)$) or a projection, otherwise f can be given as composition or differential recursion of functions from DAnalog; [4]
3. in every step of the construction of f, each component of f cannot grow faster than a quasipolynomial.

Of course, the last condition implies the first one. In this paper, we are concentrated on the restrictions $e^{\log(x)^k}$, but let us stress that we treat the Laplace transform as a fundamental tool in our work, while restrictions can be done in many different ways. Indeed, we consider that conditions 1 and 3 alone induce the class of feasible analog computations.

It is obvious to observe that DAnalog is closed under differentiation. Let us add that DAnalog is closed for integration too. We have

$\phi_1(x) = \int_0^x f(\xi) d\xi > \phi_0(x) = f(x)$ Integrating successively we get a infinite non-collapsing chain $\phi_0(x) < \phi_1(x) < \cdots < \phi_n(x) < \cdots$, where, for each pair of functions, the inequality holds almost everywhere: e.g., starting with 1, we will get all the powers n^k, for all k. We also know that the exponential is a fixed point in this hierarchy

$$e^x = \int_0^x e^\xi \, d\xi.$$

However, starting from 1 we will never cover the whole gap between 1 and e^x. If some subexponential function f is reached, then the next step will be a function γ_1 such that $\gamma_1(x) < xf(x)$. From here, we get $\gamma_2(x) < x^2 f(x),...$ and so on, a new sub-hierarchy obtained by means of polynomial factors. This means inter alia that from a polynomial we will never reach a quasipolynomial by successive integrations and that from a quasi-polynomial we will never reach a function not belonging to this class.

The property of DAnalog of being closed under differentiation and integration is considered a requirement of a well-defined analog class.

We define an admissible bounded quantification. We start with a class FUN of functions; $\exists$ FUN is defined as in the classical framework: a function $f : \mathbb{R} \to \mathbb{R}$ is in $\exists$ FUN if there exists a function $F : \mathbb{R}^{n+1} \to \mathbb{R}$ in FUN and polynomial $\phi : \mathbb{N} \to \mathbb{N}$, such that (a) $\langle x_1, \cdots, x_n \rangle \in \text{dom}(f)$ if and only if there exists a number k such that $|k| \le \phi(\lfloor x_1 \rfloor, ..., \lfloor x_n \rfloor)$ and $\langle x_1, ..., x_n \rangle \in \text{dom}(F)$ and (b) $f(x_1, ..., x_n)$ is defined and $f(x_1, ..., x_n) = y$ if and only if there exists a number k such that , $|k| \le \phi(\lfloor x_1 \rfloor, ..., \lfloor x_n \rfloor), F(x_1, ... x_n, k)$ is defined and $F(x_1, ..., x_n, k) = y$, and, for all such $|k| \le \phi(\lfloor x_1 \rfloor, ..., \lfloor x_1 \rfloor)$ such that $\langle x_1, ..., x_n, k \rangle \in \text{dom}(F)$, we have $F(x_1, ..., x_n, k) = y$.

Definition 24. The class of real recursive functions, designated by NAnalog, is defined as the collection of all real recursive functions obtained from functions in DAnalog by admissible bounded quantification (i.e. NAnalog = $\exists$ DAnalog).

Using the same techniques we used in [11] to represent by means of real recursive functions the entire arithmetical and analytical hierarchies, we can define the class NAnalog by analytical means, without quantifiers.

We get the immediate result:

Proposition 25. DAnalog ⊆ NAnalog.

Proof. This proof follows directly from definitions of DAnalog and NAnalog. Each function f of arity n in DAnalog can be seen as a function F of arity (n + 1) in DAnalog where the additional variable is introduced by composition with (n + 1)-ary projections.

Example 26. The functions $x + y$, xy, $x - y$, $\dfrac{1}{x + \varepsilon}$, for all $\varepsilon \in \mathbb{R}^+$, and

$\dfrac{x}{x + \varepsilon}$, for all $\varepsilon \in \mathbb{R}$, belong to DAnalog. The reference functions $e^{\log(x)}$, x^y, for all $k \in N$, is also in DAnalog. Also sin and cos are in DAnalog.

We can give immediately a negative result that the function e^x is not in the class DAnalog. We can add that it is possible to prove that many physical systems are in DAnalog, e.g.: RC circuits, RLC circuits, harmonic oscillators. Now, we continue the characterization of analog classes of feasible analog computation.

Proposition 27. For every function f ∈ PF, the Laplace transform L[f̂] of its canonical extension f̂ is subexponential.

Proof. Let us consider a function f of arity 1 from PF. From the definition of the class PF, the canonical extension of such a function grows as fast as $e^{\log(x)^k}$, for some positive integer k, up to a multiplicative constant. We then have, with $y = \log(x)$, $dx = e^y \, dy$, and considering without loss

of generality that k is odd, that L[f̂] (s) is given by $2 \int_0^\infty e^{-se^y + y^k + 1} \, dy$. For all $k \in \mathbb{N}$, we have

$$L[\hat{f}](s) = 2 \int_0^\infty e^{-se^y + y^k + 1} \, dy$$

and, for y greater than some y_0,

$$e^{-se^y + y^k + 1} < e^{-sy}.$$

Let $M = 2 \int_0^{y_0} e^{-se^y + y^k + 1} dy$. We have that M is defined and that

$\int_{y_0}^{\infty} e^{-sy} dy = \dfrac{e^{-sy_0}}{s}$ is defined for all s.

We may modify the proof of Proposition 15 to deal with the most interesting cases immediately.

Proposition 28. For every function f ∈ PF of arity k, there exists a real recursive function $\hat{f}$ ∈ DAnalog, of the same arity, such that f $(n_1,...,n_k)$ = $\hat{f}$ $(n_1,...,n_k)$, for every $n_1, ..., n_k$ ∈ ℕ.

Proof. The proof is given by structural induction. Let us consider the structure of functions of lower arity in PF. If the function is Z, S, I_n^j , for some n > 0 and $1 \le j \le n, \lambda x.\ 2x, \lambda x.2x + 1, \lambda x. \left| \dfrac{x}{2} \right|$, or δ, then it has its counterpart in DAnalog—this fact can easily be shown (cf. [11], where all these constructions are presented). Similar reasoning applies to basic functions of coding and decoding pairs. The definition by cases is trivial to bound.

Let h(n) = f (g(n)), where for f and g in PF we have by induction extensions $\hat{f}$ and $\hat{g}$ in DAnalog. Then $\hat{h}(x)$ = $\hat{f}$ $(\hat{g}(x))$ is an adequate representation of h. The last step is devoted to polynomially bounded recursion. Let f be given by h(x) = v (x, p(|x|)), with

$$v(x, 0) = f(x), \quad v(x, y + 1) = g(x, y, v(x, y)),$$

where for $y \leq p(|x|)$, we have $|\upsilon(x,y)| \leq q(|x|)$, where p and q are integer polynomials and $|x|$ the length of x, given by $\lfloor \log(x+1) \rfloor$ as usual.

We know (by an inductive hypothesis) that f and g have their counterparts in DAnalog, given by $\hat{f}$ and $\hat{g}$. Primarily, we are interested in the function $\hat{\upsilon}$ given by the following iteration relation:

$\hat{\upsilon}(x,y) = 1_3^3(T(y,(x,0,\hat{f}(x))))$, where $T(n, (x, y, z)) = t^n(x, y, z)$, $t(x, y, z) = (x, y + 1, \hat{g}(x, y, z))$. We can see that $\hat{\upsilon}$, defined in this way, has the

property $\hat{\upsilon}(n,y) \leq \max(\upsilon(n,\lfloor y \rfloor), \upsilon(n,\lfloor y+1 \rfloor))$ Because υ is bounded by some polynomial on the length of y, then $\hat{\upsilon}$ will be bounded by $e^{\log(x)^k}$, for some $k \in \mathbb{N}$, up to a multiplicative constant. This means that

$\hat{h}(x) = \hat{\upsilon}(x,p(\log(x)))$ is in DAnalog.

Now by virtue of Definition 24 of NAnalog, we have for free the following result:

Proposition 29. For every function $f \in$ NPF of arity k, there exists a real recursive function $\hat{f} \in$ NAnalog, of the same arity, such that $f(n_1,...,n_k) = \hat{f}(n_1,...,n_k)$, for every $n_1, ..., n \in \mathbb{N}$.

Now we change the direction of our consideration. We start from real functions and then we restrict them to the set of non-negative integers.

Definition 30. A recursive function $f : \mathbb{N} \to \mathbb{N}$ is said to be an admissible restriction of $F : \mathbb{R} \to \mathbb{R}$ if there exists an indexed ordered set $\{\alpha_i\}_{i \in \mathbb{N}}$ such that $f(i) = F(\alpha_i)$, for all $i \in \mathbb{N}$.

Then the set $\{\alpha_i\}_{i \in \mathbb{N}}$ is said to be admissible for the function F.

We are interested in admissible bounded indexed ordered sets in the sense that any two contiguous values, α_i and α_{i+1}, are bounded by a polynomial ψ, i.e., $\lfloor \alpha_{i+1} \rfloor < \psi(\lfloor \alpha_i \rfloor)$.

We can consider functions with many variables. A tuple of indexed ordered sets of real numbers   is said to be admissible for a function $F : \mathbb{R}^k \to \mathbb{R}$, of arity k, if $F(\alpha^1_{i_1}, \ldots, \alpha^k_{i_k}) \in \mathbb{N}$, for all $i_1, \ldots, i_k \in \mathbb{N}$, all sequences of real numbers are admissible bounded sets, and all pairs of indexed sets $\{\alpha^\mu_i\}_{i \in \mathbb{N}}, \{\alpha^\nu_i\}_{i \in \mathbb{N}}$ are polynomial bounded: there exists polynomial ϕ such that $\left\lVert \alpha^\mu_i \right\rVert \le \phi(\left\lVert \alpha^\nu_i \right\rVert)$ and $\left\lVert \alpha^\nu_i \right\rVert \le \phi(\left\lVert \alpha^\mu_i \right\rVert)$.

Let us present for the example that for e^x with a sequence $a_n = 2 \log n$, $n \ge 1$, we have the admissible restriction $F(n) = n^2$. Not all functions have admissible restrictions, like $\lambda x . e^{-x},$, or just like a constant $\frac{1}{2}$. Real recursive functions that extend functions over the integers have infinite admissible restrictions. For those functions f that have no admissible restrictions we can construct $\lfloor f \rfloor$, which always has admissible restrictions.

We can now formulate and prove the following proposition:

Proposition 31. Let $H : \mathbb{R}^k \to \mathbb{R}$ be a real recursive function in DAnalog. If an admissible restriction h of H, induced by an admissible set, exists, then h is in PF.

Proof. Let us consider for simplicity the case of one variable. The statement will be proved by induction in the structure of H. If H is one of the basic functions in DAnalog, then its restriction to non-negative integers (e.g., the admissible set $\{n\}_{n \in \mathbb{N}}$) is obviously in PF.

The last part of the proof is devoted to differential recursion. Consider a function H of arity one, defined by $H(0) = F (= 0$, without loss of generality), $\partial_y H(y) = G(y, H(y))$ such that $H(y) \le e^{\log(y)^k}$, for some k, up to a multiplicative constant. Let α be an admissible set for H. Of course we can rearrange _i to be strictly monotonic. Then $H(\xi) = \int_0^\xi G(y, H(y)) dy$. We can divide the right-hand side of this equation into intervals $[\alpha_i, \alpha_{i+1}]$, and use the mean value theorem to obtain $H(\alpha_n) = \sum_{i=0}^{n-1} G(\xi_i, H(\xi_i))$, where $G(\xi_i, H(\xi_i)) = \int_{\alpha_i}^{\alpha_{i+1}} G(y, H(y)) dy$. Since in DAnalog there are functions from $L_{k'}$ hence the mean value theorem can be used here. If α is the admissible set for H, then $H(\alpha_i)$ can be presented as a sum of values of some natural function k(i). For $H(\alpha_i) \in$ $\mathbb{N}$, we have $\sum_{i=0}^{n-1} G(\xi_i, H(\xi_i)) = H(\alpha_i) \in \mathbb{N}$. This holds for all $n > 0$, hence $G(\xi_i, H(\xi_i)) \in \mathbb{N}$, for every $i \in \mathbb{N}$. Let $\gamma_i = H(\xi_i)$. We define $k(i) = G(\xi_i, \gamma_i)$. From an inductive hypothesis k(i) is in PF, then we find that h is in

PF:Gis polynomially bounded and $h(i) = \sum_{j=0}^{i-1} k(j)$, and the sum is polynomially bounded by the condition on G, hence h is also in PF.

Let the reader see how the last proof is based on an analytic result—the mean value theorem. We stress this fact to show how the intermixing computability and analysis provide an analytic language to the theory of computation. Approximation techniques can also be used like in Computable Analysis (see, e.g., [17]). However, these approximation techniques give rise to differential equations having solutions with large plateaux between consecutive integers and fast growing rates close to the integer values of the argument. Similarly, we can prove that:

Proposition 32. Let $H: \mathbb{R}^k \to \mathbb{R}$ be a real recursive function in NAnalog. If an admissible restriction h of H, induced by an admissible polynomial bounded set, exists, then h is in NPF.

Proof. Every function f of arity n in NAnalog corresponds to a function F of arity n + 1 in DAnalog. Since f comes from F by universal quantification in the last variable of F, all admissible restrictions of f are extensible to admissible restrictions of F. According to Proposition 31, all these extended admissible restrictions, say ζ, lie in PF. But any such ζ is suitable for quantification in the last variable, providing functions in NPF. Reasoning this way, we can also suppose that all functions involved grow slower than $\lambda x. e^{\log(x)^k}$ in all its arguments.

THE CONJECTURE DANALOG ≠ NANALOG

We now show that the nature of computational classes PF and NPF induces a relation (strict inclusion or equality) between analog classes DAnalog and NAnalog, respectively, analogous to the classical open relation between P and NP

Let us indicate the following property, which is implied by Propositions 28 and 29: if f ∈ PF or f ∈ NPF then its canonical extension F is bounded by a quasi-polynomial in any step of the construction of F.

If we restrict classes DAnalog and NAnalog to classes of functions that have admissible restrictions—and this will not be loosing generality, but rather a definition strategy that can be restated as we saw before in Section 4—then Proposition 34 and the main theorem of this section will follow.

Proposition 33. If a function has an admissible restriction in any step of its construction, and in any step of its construction all its admissible restrictions are quasi-polynomially bounded, then it belongs to DAnalog.

Proof. Suppose that a function f of arity one in these conditions is not in DAnalog. Then we can find, in some step of the construction a component g such that, for all $k \in \mathbb{N}$, $g(x) \geq 2^{\log(x)^k}$. The function g has an admissible restriction which is not quasi-polynomially bounded, contrarily to our hypothesis. Thus f is in DAnalog.

Proposition 34. If NP ⊆ P then NAnalog ⊆ DAnalog.

Proof. If f is a function in NAnalog, then all their admissible restrictions along its construction are non-deterministic quasi-polynomially bounded (i.e., they are in NPF), and then, by hypothesis, they are deterministic quasi-polynomially bounded (i.e., they are in PF). We end the proof applying Proposition 33.

And now, by the simple contraposition we can finally state the main result.

Proposition 35. If NAnalog ≠ DAnalog then NP ≠ P.

Hence, considering quasi-polynomially bounded functions over the real numbers, which have admissible restrictions, then an analytic proof that DAnalog ≠ NAnalog (in the continuum) will provide a proof that P is a proper subclass of NP.

We now are in a proper situation to ask if there is simple mathematical condition in Complex Analysis to check the conjecture DAnalog ≠ NAnalog. That is, to express the conjecture P ≠ NP and solve it by means of analytical tools. The precedent proposition implies that P ≠ NP if, for some function f ∈ NAnalog, which has admissible restrictions, does not exist a new system of differential equations for f in DAnalog.

CONCLUSIONS

In this paper we presented formally only the implication: if NAnalog ≠ DAnalog then NP ≠ P. We leave to the reader the proof of the converse implication: Let f ∈ NPF; then f has a canonical extension in NAnalog; if NAnalog ≠ DAnalog, then this canonical extension is also in DAnalog; we conclude that the function f is in PF as we wished to show.

We are now considering Sobolev spaces. We wanted to show here that the consideration of feasible analog circuits made of elastic strings and viscous dashpots [7] lead to the Laplace transform as a measure of feasible analog computation. Away of physical considerations, the use of Sobolev spaces will allow the replacement of the Laplace transform by a suitable transform measuring rates of growth adapted to our cases. This direction of research is, at it seems, reasonable and, possibly, most fruitful for a better representation of P in the analog realm.

As usual, we cannot close this paper without listing some still open problems:

- Is REC($\mathbb{C}$) richer than REC($\mathbb{R}$) under growth constraints? E.g., is the class DAnalog over $\mathbb{C}$ richer than the class DAnalog over $\mathbb{R}$? We believe that the answer is yes, with regard to the different power of Cauchy's problem in real and complex space.

- The simulation of a non-deterministic Turing machine designed to compute a function shows clearly that NAnalog is well conceived: a suitable guess exists such that the value of the function, for a given input, is computed. Can this function be obtained by integration in the complex plane? We think that a deep exploration of this idea would bring more light to P ≠ NP conjecture.
- Can the Laplace transformation pair replace the differential recursion scheme in the case of some computational subclasses of real recursive functions? That would be a quite elegant formulation.

ACKNOWLEDGEMENTS

Thanks to Agnieszka for her support. We thank Samson Abramsky from CompLab in Oxford and Nachum Dershowitz from the School of Computer Science, Tel Aviv University, for their encouragement. Martin Davis was precious to us, whether being encouraging or critical. Martin namque erit ille mihi semper deus. [5] Our friend Professor António Portela from the Technical University of Lisbon pointed to us very interesting physical aspects of the Laplace transform in dissipative systems, being of a mechanical nature, an electric nature, whenever friction exists. Part of Section 2 was produced as an answer to many questions raised by Fernando Ferreira from the University of Lisbon.

We also thank the two referees who helped to make this version of the paper. To Cris Moore, from New Mexico University, we express our gratitude and admiration for having been the mentor of our work since 1995.

REFERENCES

1. J.L. Balcázar, J. Díaz, J. Gabarró, Structural Complexity I, second ed., Springer, Berlin, 1995.
2. O. Bournez, E. Hainry, Real recursive functions and real extensions of recursive functions, Machines, Computation, Universality (MCU 2004), Lecture Notes in Computer Science, vol. 3354, 2005, Springer, Berlin, pp. 116–127.
3. M.L. Campagnolo, Continuous-time computation with restricted integration capabilities, Theoret. Comput. Sci. 317 (2004) 147–165.
4. M.L. Campagnolo, C. Moore, J.F. Costa, Iteration, inequalities, and differentiability in analog computers, J. Complexity 16 (4) (2000) 642–660.
5. M.L. Campagnolo, C. Moore, J.F. Costa, An analog characterization of the Grzegorczyk hierarchy, J. Complexity 18 (4) (2002) 977–1000.

6. D. Graça, J.F. Costa, Analog computers and recursive functions over the reals, J. Complexity 19 (5) (2003) 644–664.
7. D.E. Hyndman, Analog and Hybrid Computing, Pergamon Press, New York, 1970.
8. P. Koiran, C. Moore, Closed-form analytic maps in one and two dimensions can simulate universal Turing machines, Theoret. Comput. Sci. 210 (1) (1999) 217–223.
9. C. Moore, Recursion theory on the reals and continuous-time computation, Theoret. Comput. Sci. 162 (1996) 23–44.
10. J. Mycka, -recursion and infinite limits, Theoret. Comput. Sci. 302 (2003) 123–133.
11. J. Mycka, J.F. Costa, Real recursive functions and their hierarchy, J. Complexity 20 (6) (2004) 835–857.
12. J. Mycka, J.F. Costa, The computational power of continuous dynamic systems. Machines, Computation, Universality (MCU 2004), Lecture Notes in Computer Science, vol. 3354, 2005, Springer, Berlin, pp. 164–175.
13. J. Mycka, J.F. Costa, Undecidability over continuous-time, Logic J. IGPL, Oxford University Press, in print.
14. P. Odifreddi, Classical Recursion Theory I, Elsevier, Amsterdam, 1992.
15. P. Odifreddi, Classical Recursion Theory II, Elsevier, Amsterdam, 1999.
16. A. Vretblad, Fourier Analysis and Its Applications, Graduate Texts in Mathematics, vol. 223, Springer, Berlin, 2003.
17. K. Weihrauch, Computable Analysis, An Introduction, Texts in Theoretical Computer Science, Springer, Berlin, 2000.

CITATION

Jerzy Mycka and José Félix Costa, The P≠NP Conjecture in the Context of Real and Complex Analysis, doi:10.1016/j.jco.2005.07.003.

Primitive Recursiveness of Real Numbers under Different Representations

Qingliang Chen[a, b,] Kaile Sua, [c,] and Xizhong Zhengb, [d,]

[a] Department of Computer Science, Sun Yat-sen University, Guangzhou 510275, P.R.China
[b] Theoretische Informatik, BTU Cottbus, Cottbus 03044, Germany
[c] Institute for Integrated and Intelligent Systems, Griffith University, Brisbane, Qld 4111, Australia
[d] Department of Computer Science, Jiangsu University, Zhenjiang 212013, P.R.China

ABSTRACT

In mathematics, various representations of real numbers have been investigated. All these representations are mathematically equivalent because they lead to the same real structure—Dedekind-complete ordered field. Even the effective versions of these representations are equivalent in the sense that they define the same notion of computability of real numbers. However, the primitive recursive (p.r., for short) versions of these representations can lead to different notions of p.r. real numbers. Several interesting results about p.r. real numbers can be found in literatures. In this paper we summarize the known results about the primitive recursiveness of real numbers for different representations as well as show some new relationships. Our goal is to clarify systematically how the primitive recursiveness depends on the representations of the real numbers.

INTRODUCTION

The computability of real numbers is introduced by Alan Turing in his seminal paper [19]. According to Turing, "the 'computable' numbers may be described briefly as the real numbers whose expressions as a decimal are calculable by finite means". In order to define the "finite means" precisely, he introduces the nowadays well-known Turing machines. Since Turing machines compute exactly the computable functions on natural numbers, Turing defines actually the real numbers with computable decimal expansions as computable real numbers. Namely, x is computable if there is a computable function

$f: \mathbb{N} \rightarrow \{0, 1, \cdots, 9\}$ such that $x = \sum_{n=0}^{\infty} f(n).^{-(n-1)}$ Here we consider only the real numbers in the interval $[0, 1]$. As it was pointed out by Robinson [16], Myhill [11], Rice [15] and others, the computability of real numbers can be equivalently defined by means of Cauchy sequences, Dedekind cuts and other representations of real numbers. That is, the computability of reals is independent of their representations. The class of computable reals will be denoted by EC (for Effectively Computable).

Besides the computability, the sub recursive real numbers like primitive recursive and polynomial time computable real numbers have also been discussed. The different notions of sub recursive real numbers could be defined if different representations are used. Specker [18] is the first who investigates this problem and he shows that decimal expansions, Dedekind cuts and Cauchy sequences lead to three different versions of p.r. real numbers. Later on, Peter [14], Mostowski [10], and Lehman [9] investigated other versions of p.r. reals and showed some more relations between the notions of p.r. real numbers based on different representations. However, not every important representation of real numbers have been discussed and there is no a systematically overview about the subrecursiveness of real numbers so far.

This paper aims to address the deficit. We summarize the known results about primitive recursiveness of reals which we can find in literatures. We will give some new properties of the p.r. reals and analyse systematically the dependence of primitive recursiveness of reals on the representations. This paper is organized as follows. Firstly we recall the representations in the computability theory for real numbers in the next section, and then we will survey and explore the hierarchy in these representations in section 3, 4, 5, 6 and 7 by nested intervals, Cauchy sequences, b-adic expansion, Dekedind cut and continued fraction, respectively. And we will conclude the paper in the last section.

REPRESENTATIONS OF REAL NUMBERS

In this section, we recall the representations of real numbers which will be discussed in this paper. First we explain the classical form of the representations. Since we are interested in the effectivizations of the representations to different levels, all representations will be defined again in a uniform way such that they depend on some given class $\mathcal{F}$ of functions. According to the choice of the class $\mathcal{F}$, various computability of different levels about real numbers can be defined. These notions depend also on the selected representations.

For simplicity, we consider only the real numbers in the unity interval $[0, 1]$. If a real number x is not in this interval, then there is a $y \in [0, 1]$ and a natural number n such that $x = y + n$ or $x = y - n$. In this case, the real numbers x and y should have the same computability level in any reasonable sense.

Now we recall the representations of real numbers informally

A sequence (x_s) of rational numbers is called a Cauchy sequence if, for any $\in > 0$, there is an N such that $|x_s - x_t| \leq \in$ for all s, t $\geq$ N. That is, Cauchy sequences are simply the converging sequences. A Cauchy sequence (x_s) represents a real number x if the sequence converges to x. This representation is called naive Cauchy representation (see Weihrauch [20]). In other words, a naive Cauchy representation of a real number x is a sequence (x_s) of rational numbers which converges to x. A more popular representation by Cauchy sequence in computable analysis uses the Cauchy sequence with an effective convergence modulus and we call this representation simply Cauchy representation. More precisely, a Cauchy representation of a real number x is a sequence (x_s) of rational numbers which converges to x effectively in the sense that $|x_s - x| \leq 2^{-s}$ for all s. Some variations of Cauchy representation will be discussed in the section 4.

A Dedekind cut is a pair (C, D) of sets of rational numbers such that C is closed downward, i.e., if $u \leq v$ and $v \in C$ then $u \in C$, and D is upward, i.e., if $u \leq v$ and $u \in D$ then $v \in D$. A Dedekind cut (C, D) represents a real number x means actually that x is the least upper bound of C. Since a Dedekind cut (C, D) is uniquely determined by the set C, we define usually the (left) Dedekind cut of x as the set $C_x := \{r \in \mathbb{Q} : r < x\}$ of rational numbers and regard the set C_x as the Dedekind cut representation of x. Since any set can be described uniquely by its characteristic function, we can also define the Dedekind cut representation of a real number x as a function $f\colon \mathbb{N}^2 \to \{0, 1\}$ such that $f(n, m) = 1$ if and only if $n/m < x$.

The decimal representation might be the most well-known representation of real numbers. If x is a real number in the interval $[0, 1]$, then x can be denoted by a decimal expansion $x = 0.a0a_1a_2a_3 \cdots$ where $a_s \in \{0,$ 1... 9$\}$ such that $x = \sum_{s=0}^{\infty} a_s \cdot 10^{-(s+1)}$. The sequence (as) corresponds to a function $f\colon \mathbb{N} \to \{0, 1... 9\}$. Thus we can define the decimal representation of a real number $x \in [0, 1]$ as a function $f\colon \mathbb{N} \to \{0, 1... 9\}$ such that $x = \sum_{s=0}^{\infty} f(s) 10^{-(s+1)}$. The decimal representation represents the real numbers in base 10. In general, for any natural number $b > 1$, we can also represent real numbers in base b. This is the b-adic expansion. That is, a b-adic representation of a real number x is a function $f\colon \mathbb{N} \to \{0, 1,...,b - 1\}$ such that $x = \sum_{s=0}^{\infty} f(s) \cdot b^{-(s+1)}$. If the base $b = 2$, then the b-adic representation is called binary representation.

The binary representation of real numbers relates a real number to a set of natural numbers in a very natural way. Let $f\colon \mathbb{N} \to \{0, 1\}$ be a binary representation of a real number x. Then we have $x = \sum_{s=0}^{\infty} f(s)\, 2^{-(s+1)}) = \sum_{s \in A} 2^{-(s+1)}$ where $A := \{s \in \mathbb{N} : f(s) = 1\}$. Thus the binary representation is the characteristic function of A. The real number x is usually denoted also by $x = x_A$.

Another representation of real numbers is by sequences of nested rational intervals. A sequence (I_s) of closed intervals with rational endpoints is called nested if $I_{s+1} \subseteq I_s$ for all s. This sequence represents a real number x, if x is the unique common member of all intervals. The interval sequence can be defined by two functions. Therefore, we can define the nested interval representation of a real number x as a function pair f, g: $\mathbb{N} \to \mathbb{Q}$ such that $f(s) \leq f(s + 1) \leq x \leq g(s + 1) \leq g(s)$ for all s and $\lim_{s \to \infty} (g(s) - f(s)) = 0$. Finally, we explain the continued fraction expansion of real numbers. It is known that every positive real number x has a unique regular continued fraction expansion of the form

$$x = b_0 + \cfrac{1}{b_1 + \cfrac{1}{b_2 + \dots}}$$

(1)

Where $b_0 \geq 0$, $b_n \geq 1$ for $n \in \mathbb{N}$. For brevity, we write $x = [b_0, b_1, b_2, \dots]$ in which the number bn is called the partial quotient of order n. It is obvious that in the case for the rational numbers, the numbers of b_n is finite. That is, if x is rational, then $x = [b_0, b_1, b_2, \dots, b_n]$ for some n. For convenience, we denote this rational number by $x = [b_0, b_1, b_2, \dots, b_n, 0, 0, \dots]$. Thus, a continued fraction representation of a real number x is a function f : $\mathbb{N} \to \mathbb{N}$ such that $x = [f(0), f(1), f(2), \dots]$.

All representations mentioned above are mathematically equivalent because they deduce the same (more precisely, isomorphic) structure which is called Dedekind-complete ordered field. However, if we are interested in the computability of real numbers, the situation is different. In order to explain this more precisely, we look at firstly how the computability notion can be introduced to real numbers. The idea is very simple. As we have seen, every representation of real numbers mentioned above uses functions, either from natural numbers to natural numbers or from natural numbers to rational numbers. Thus, any effectively notion about these functions can be transferred naturally to the real numbers represented by these functions. To this end, we have to extend the computability (or sub computability) of the functions on natural numbers to the functions from natural numbers to rational numbers. This extension looks like the following:

Let $\mathcal{F}$ be a class of some functions f: $\mathbb{N} \to \mathbb{N}$. We say that a function g: $\mathbb{N} \to \mathbb{Q}$ belongs to $\mathcal{F}$ means that there are functions a, b, c: $\mathbb{N} \to \mathbb{N}$ in $\mathcal{F}$ such that g(n)=(a(n) − b(n))/(c(n) + 1) for all n.

Now we give the precise definition of the relativization of all above representations to a class $\mathcal{F}$ of functions.

Definition 2.1: Let $\mathcal{F}$ be a class of functions f: $\mathbb{N} \to \mathbb{N}$ or f: $\mathbb{N} \to \mathbb{Q}$, and let x ∈ [0, 1] be a real number.

(i) x has an $\mathcal{F}$-Cauchy representation (x ∈ CS($\mathcal{F}$)) if there is a function f: $\mathbb{N} \to \mathbb{Q}$ in F such that the sequence f(s) converges to x effectively.

(ii) x has an $\mathcal{F}$-Dedekind cut representation (x ∈ DC($\mathcal{F}$)) if there is a function f : $\mathbb{N}^2 \to \{0, 1\}$ in $\mathcal{F}$ such that f(n, m) = 1 if and only if n/(m + 1) < x

(iii) x has a b-adic representation (x ∈ bAE($\mathcal{F}$)) if there is a function f : $\mathbb{N} \to \{0, 1, \cdots, b - 1\}$ in $\mathcal{F}$ such that $x = \sum_{s=0}^{\infty} f(s) \cdot b^{-(s+1)}$. Especially, for b = 10 and b = 2, they are a decimal and binary representation, respectively.

(iv) x has a continued fraction representation (x ∈ CF ($\mathcal{F}$)) if there is a function f: $\mathbb{N} \to \mathbb{N}$ in $\mathcal{F}$ such that x = [f (0), f (1), ⋯].

(v) x has a nested interval representation (x ∈ NI($\mathcal{F}$)) if there are two functions f, g: $\mathbb{N} \to \mathbb{Q}$ in F such that f(s) ≤ f(s + 1) ≤ x ≤ g(s + 1) ≤ g(s) for all s and lims→∞(g(s) − f(s)) = 0.

When we limit the function class $\mathcal{F}$ to be p.r. functions, it will lead to the definitions of various versions of "p.r. real numbers". Denote by R_4, R_3, R_2^b, R_1 and R_0 the classes of real numbers which have p.r. continued fraction, p.r. Dedekind cut, p.r. b-adic expansion, p.r. Cauchy representation, and p.r. nested interval representations, respectively. That is,

$$R_4 = CF(\mathcal{F}), R_3 = DC(\mathcal{F}), R_2^b = bAE(\mathcal{F}), R_1 = CS(\mathcal{F}), R_0 = NI(\mathcal{F}).$$

We will see that the relationship among these classes is as follows.

$$R_4 \subsetneq R_3 \subsetneq R_2^b \subsetneq R_1 \subsetneq R_0 = EC$$

THE NESTED INTERVAL REPRESENTATION

In this section we discuss the representation of reals by p.r. nested interval. And we will see that a real number has a p.r. nested interval representation if and only if it is computable.

By definition, a p.r. nested interval representation of a real number x supplies the p.r. upper and lower bounds of x to any precision. This is equivalent to a p.r. approximation to x with a p.r. error estimation which is called a p.r. approximation of x by Skordev [17]. More precisely, a p.r. approximation of a real x is a pair (a, e) of p.r. functions a, e: $\mathbb{N} \to \mathbb{Q}$ such that

(i) e is monotonically decreasing and converges to 0; and

(ii) $|a(n) - x| \le e(n)$ holds for all $n \in \mathbb{N}$.

Lemma 3.1: A real number has a p.r. nested interval representation if and only if it has a p.r. approximation.

Proof: Suppose that x has a p.r. nested interval representation (f, g). That is, for any n, we have $f(n) \le f(n + 1) \le x \le g(n + 1) \le g(n)$ and $\lim_{n \to \infty}(g(n) - f(n)) = 0$. Define two p.r. functions a, e : $\mathbb{N} \to \mathbb{Q}$ by $a(n)=(g(n) + f(n))/2$ and $e(n)=(g(n) - f(n))/2$ for all n. Then (a, e) is a p.r. approximation of x.

On the other hand, if x has a p.r. approximation (a, e), then a (n) −e (n) ≤ x ≤ a(n) + e(n) for all n ∈ ℕ. We can define two p.r. functions f and g by f(n) := max{a(t) − e(t) : t ≤ n} and g(n) := min{a(t) + e(t) : t ≤ n}. The function pair (f, g) is obviously a p.r. nested interval representation of x.

Remember that any computable number x has a computable approximation with computable error estimation. The next result shows that computable real numbers have even p.r. approximations.

Theorem 3.2 (Skordev [17]): A real number is computable if and only if it has a p.r. approximation.

Proof: If a real number x has a p.r. approximation (a, e), then, for all n, we have |a(s(n)) − x| ≤ 2−n for the computable function s defined by s(n) = µi(e(i) ≤ 2−n). That is, x is computable

From the other direction, if x is computable, then there is a computable function f: ℕ → ℚ such that |f (n) − x| ≤ 2^{-n}. Let M be a Turing machine which computes the function f. According to the Kleene's predicate [5], there is a p.r. predicate T such that T (n, y, s) holds if and only if the machine M with the input n outputs y in s steps. Therefore, f (n) = y if and only if T (n, y, s) for some s.

Fix an a ∈ ℕ and b ∈ ℚ such that f (a) = b and define two p.r. functions g, h: ℕ → ℕ by

$$g(\langle n, y, s \rangle) = \begin{cases} n & \text{if } T(n, y, s); \\ a & \text{otherwise.} \end{cases}$$

$$h(\langle n, y, s \rangle) = \begin{cases} y & \text{if } T(n, y, s); \\ b & \text{otherwise.} \end{cases}$$

Here $\langle .,.,. \rangle$: ℕ × ℚ × ℕ → ℕ is a p.r. pairing function. It is easy to see that h (n) = f (g(n)) holds for all n.

Let g be an unbounded monotonically increasing primitive function de- fined by

$$g'(n) = \max\{g(i) : 1 \le i \le n\}.$$

Then the primitive function

$$h'(n) = f(g'(n)).$$

Satisfies

$$|h'(n) - x| = |f(g'(n)) - x| \le 2^{-g'(n)}.$$

So $(h'(n), 2^{-g'(n)})$

is a p.r. approximation for x.

Corollary 3.3: A real number has a p.r. nested interval representation if and only if it is computable. That is, $R_0 = E_C$.

CAUCHY REPRESENTATION

Although the primitive recursive approximation (a, e) of a real number x mentioned in the Section 3 has a p.r. error estimation e, this estima- tion converges to 0 not necessarily fast enough. For example, it is not guaranteed that, for any n, a stage m can be found such that e (m) ≤ 1/n. That is the reason, why the real numbers of p.r. approximations does not form a proper subset of computable real numbers. In order to introduce properly the notion of primitive recursive real number, we should require that also the error estimation converges to 0 primitive recursively. The p.r. Cauchy representation discussed in this section supplies a good approach to this direction.

As mentioned in the Section 2, the p.r. Cauchy representation of a real number x is simply a p.r. function f: $\mathbb{N} \to \mathbb{Q}$ such that $|x - f(n)| \leq 2^{-n}$. That is, it is a p.r. approximation with the error estimation function e $(n) := 2^{-n}$. The choice of the function $\lambda n.2^{-n}$ is not essential. The function $\lambda n.n^{-1}$ is also widely used in literature. Actually, and p.r. function e: $\mathbb{N} \to \mathbb{Q}$ which converges to 0 primitive recursively suffices. In this paper, we use either the function $\lambda n.2^{-n}$ or $\lambda n.n^{-1}$ without further explanation.

Similar to the case of the computable Cauchy representation, not only the error estimation function has different choices, p.r. Cauchy representation of a real number can also be stated in several different but equivalent ways.

Proposition 4.1: Let x be a real number. Then the following conditions are equivalent.

(i) x has a p.r. Cauchy representation f, i.e., $|x - f(n)| \leq 2^{-n}$ for all n;

(ii) There is a p.r. function f: $\mathbb{N} \to \mathbb{Q}$ and a p.r. function e: $\mathbb{N} \to \mathbb{N}$ such that

$$(\forall n, m \in \mathbb{N})(m \geq e(n) \implies |f(m) - x| \leq 2^{-n}). \tag{2}$$

(iii) There is a p.r. function f: $\mathbb{N} \to \mathbb{Q}$ such that $\lim_{n \to \infty} f(n) = x$ and

$$(\forall n, m \in \mathbb{N})(n \leq m \implies |f(n) - f(m)| \leq 2^{-n}). \tag{3}$$

(iv) There is a p.r. function f: $\mathbb{N} \to \mathbb{Q}$ and a p.r. function e: $\mathbb{N} \to \mathbb{N}$ such that $\lim_{n \to \infty} f(n) = x$ and

$$(\forall n, m \in \mathbb{N})(n, m \geq e(k) \implies |f(n) - f(m)| \leq 2^{-k}). \tag{4}$$

Proof: "(i) $\Rightarrow$ (ii)" It suffices to define e (n):= n for all n. "(ii) $\Rightarrow$ (iii)" Let f and e be a p.r. functions which satisfy the condition

(2). Define a p.r. function f_1 (n):= f (e (n + 1)) for all n. Then for any n $\leq$ m we have

$$|f_1(n) - f_1(m)| \leq |f_1(n) - x| + |f_1(m) - x|$$
$$= |f(e(n+1)) - x| + |f(e(m+1)) - x|$$
$$\leq 2^{-(n+1)} + 2^{-(m+1)} \leq 2 \cdot 2^{-(n+1)} = 2^{-n}$$

That is, the primitive function f_1 satisfies the condition (3).

"(iii) $\Rightarrow$ (vi)" Let f be a p.r. function with $\lim_{n \to \infty}$ f (n) = x which satisfies the condition (3). Define e (n):= n + 1 and f_1 (n):= f (n + 1). Then, for any n, m $\geq$ e (k) we have $|f_1$ (n) $- f_1$ (m) $|\leq|f$ (n + 1) $-$ f (k + 1)| + |f (m + 1) $-$ f (k + 1)| $\leq 2^{-k}$. That is, f1 and e satisfy (4). "(vi) $\Rightarrow$ (i)" Let f : $\mathbb{N}$ $\to \mathbb{Q}$ and e : $\mathbb{N} \to \mathbb{N}$ be p.r. functions such that $\lim_{n \to \infty}$ f(n) = x and |f(n) $-$ f(m)| $\leq 2^{-k}$ for all n, m $\geq$ e(k). Define f_1 (n):= f (e (n) + 1) for all n and let m $\to \infty$, we get $|f_1$ (n) $-$ x| $\leq 2^{-n}$ for all n. That is, f_1 is a p.r. Cauchy representation of x.

A real number x is called Cauchy p.r. if it has a primitive Cauchy representation. The class of Cauchy p.r. real numbers is denoted by R_1.

Theorem 4.2: The class of Cauchy p.r. real numbers is closed under arithmetical operations. That is, it is a field.

Proof: Let x and y be Cauchy p.r. real numbers and let f and g be p.r. Cauchy representations of x and y, respectively. Then the function h defined by h (n):= f(n + 1) + g(n + 1) is a p.r. Cauchy representation of x + y. That is, x + y is Cauchy p.r. too.

Choose a natural number k such that max {|f(n)|, |g(n)|, |x|, |y|} $\leq 2^k$ for all n. Since |f(n)g(n) $-$ xy|$\leq$|f(n)||g(n) $-$ y| + |y||f(n) $-$ x| $\leq 2^{-(n-k-1)}$, the function h defined by h(n) := f(n + k + 1)g(n + k + 1) is a p.r. Cauchy representation of xy and hence xy is also Cauchy p.r.

If y $\neq$ 0, then there is a constant k such that min {|y|, |g (n)|} $\geq 2^{-k}$ and max {|f (n)|, |g (n)|} $\leq 2^k$ for all n. Since

$$\left|\frac{f(n)}{g(n)} - \frac{x}{y}\right| = \left|\frac{f(n)y - g(n)x}{g(n)y}\right|$$

$$\leq \frac{|f(n)||g(n) - y| + |g(n)||f(n) - x|}{|g(n)y|} \leq 2^{-n+3k+1}.$$

Thus, the function h defined by h (n) = f (n + 3k + 1)/g(n + 3k + 1) is a p.r. Cauchy representation of x/y and hence x/y is Cauchy p.r.

By a simply diagonalization against all p.r. Cauchy representations, it is easy to construct a computable real number which does not have p.r. Cauchy representation. That is we have

Theorem 4.3: The class of Cauchy p.r. reals is a proper subset of computable reals. That is, R1 $\subsetneq$ EC.

REPRESENTATIONS BY B-ADIC EXPANSION

In mathematics real numbers are usually represented by its decimal expansion while in computer science the binary expansions are more popular. Of course, real numbers can also be represented in b-adic expansions for any b > 1. We call a real x b-adic primitive recursive (or b-adic p.r., in short) if there is a p.r. function f such that $x = \sum_{n=0}^{\infty} f(n) \, b^{-(n+1)}$. The class of all b-adic p.r. real numbers is denoted by R_2^b. In this section we can see that the classes R_2^b are proper subsets of the Cauchy p.r. real number class R_1 for all b > 1. Besides, for different b, the classes R_2^b are not necessarily the same.

Notice that, if x has a p.r. b-adic expansions, i.e., $x = \sum_{n=0}^{\infty} f(n) \, b^{-(n+1)}$ for a p.r. function f, then the function g defined by $g(n) = \sum_{i=0}^{\infty} f(i) \, b^{-(i+1)}$ is a p.r. Cauchy representation of x. This observation implies immediately that $R_2 \subseteq R_1$.

On the other hand, Specker [18] shows that $R_2^{10} \neq R_1$. The proof's idea of Specker for the inequality $R_2^{10} \neq R_1$ is to show that $R^{10}{}_2$ is not closed under arithmetical operations. Since R_1 is closed under arithmetical operations, they are different. This idea can be extended to the case of other base b. Thus, the primitive recursiveness of real numbers based on Cauchy representation and b-adic expansions are different.

Specker's proof uses the following technical lemma which can be proved by the Kleene normal form theorem [5] for the computable functions.

Lemma 5.1 (Specker [18]): There are p.r. functions u and v such that the function q defined by

$$q(n) := u((\mu t \geq n)(v(t) = 0))$$

is not a p.r. function.

Theorem 5.2 (Specker [18]): There exists a decimal p.r. real number x such that 3x is not decimal p.r.

Proof: We want to find a real number x: $= 0.a_0 a_1 a_2 a_3 \cdots$ such that the function a defined by a (i):= a_i is p.r., but for $3x = b.b_0 b_2 b_2 b_3 \cdots$, the function b defined by b (i):= b_i is not p.r. Let's look first at an example. For simplicity, we restrict that $a_i \in \{1, 3, 5\}$ for all i.

$$x = 0 . a_0 a_1 a_2 \cdots \quad := 0.335131155331511\cdots$$

$$3x = b . b_0 b_1 b_2 \cdots \quad := 1.00539346599453?\cdots$$

where " ? " can be 3 or 4 depending on the first non-3 digit of x after this place is equal to 1 or 5.

Check the odd-even property of b_i we can find that a digit b_i is even if and only if the first digit a_j with j>i and $a_j \neq 3$ is 5. That is, for any n, we have

$$b(n) \equiv 0 \mod 2 \iff a((\mu t > n)(a(t) \neq 3)) = 5.$$

$$(5)$$

Now the problem is reduced to find a p.r. function a: $\mathbb{N} \to \{1, 3, 5\}$ such that the function b satisfying condition (5) is not p.r... It suffices to find a p.r. function a: $\mathbb{N} \to \{1, 3, 5\}$ such that the function q: $\mathbb{N} \to \{1, 5\}$ defined by

$$q(n) := a((\mu t > n)(a(t) \neq 3))$$

$$(6)$$

Is not p.r. For any p.r. functions u: $\mathbb{N} \to \{1, 5\}$ and v: $\mathbb{N} \to \mathbb{N}$, if we define the function a by

$$a(n) := 3 \cdot sg(v(n)) + u(v(n)) \cdot \overline{sg}(v(n))$$

Where sg and $\overline{sg}$ are signal functions. Then we have a (n) $\in \{1, 3, 5\}$ and a(n) $\neq$ 3 if and only if v(n)=0& a(n) = u(n) for all n. This implies that

$$a((\mu t > n)(a(t) \neq 3)) = u((\mu t > n)(v(t) = 0))$$

Thus, the problem is reduced further to find two p.r. functions u: $\mathbb{N} \to \{1, 5\}$ and v: $\mathbb{N} \to \mathbb{N}$ such that the function defined by

$$q(n) := u((\mu t > n)(v(t) = 0))$$

is not primitive recursive. The existence of such p.r. functions u and v follows from the Lemma 5.1.

Remark 5.3: By the same kind of construction of Specker's, it is not hard to see that any p.r. b-adic expansion real is not closed under arithmetical operations.

Corollary 5.4: The class of b-adic expansions p.r. reals is a proper subset of the class of Cauchy p.r. real numbers, i.e., $R_2^b \subsetneq R_1$.

Now we explore the relationship among the classes R_2^b for different b's. Firstly we look at a simple example. Let b, d > 1 be bases such that there is a k > 0 such that $b^k = d$. If x is a b-adic p.r. real number,

i.e., x: $= \sum_{n \in \mathbb{N}} f(n) b^{-(n+1)}$ for a p.r. function f, then the function g (n)

$\sum_{i=0}^{k-1} f(nk + i)$ is also p.r. and hence $x = \sum_{n \in \mathbb{N}} g(n) d^{-(n+1)}$ is d-adic p.r. too. This observation has been extended by Mostowski [10] to the following result.

Theorem 5.5 (Mostowski [10]): Let b, d > 1. If a power of b is divisible by d, then any b-adic p.r. real is also d-adic p.r., i.e., $R_2^b \subseteq R_2^d$.

Proof: Suppose that $b^k = s\,d$ for some k, s $\in \mathbb{N}$ and x = $\sum_{n=0}^{\infty} f(n) b^{-(n+1)}$

is a p.r. b-adic expansion of x. Define a p.r. function g by g (n):$= \sum_{i=0}^{k-1} f$ (n · k + i) · b^{k-1-i}. Notice that, g (n) $\le (1 + b + \cdots + b^{k-1})(b-1) = b^k - 1$, because f(i) $\le$ b−1 for all i. Now the d-adic expansion of x can be obtained as follows.

$$x = \sum_{n=0}^{\infty} f(n)b^{-(n+1)} = \sum_{n=0}^{\infty} g(n)b^{-k(n+1)}$$

(7)

$$= \sum_{i=0}^{n} h(i)d^{-(i+1)} + r(n)b^{-k} + \sum_{i=n}^{\infty} g(i)b^{-k(i+1)} = \sum_{n=0}^{\infty} h(n)d^{-(n+1)}$$

(8)

Where the p.r. functions r and h are defined inductively by

$$\begin{cases} h(0) := qt(g(0), s) \\ r(0) := rs(g(0), s) \\ h(n+1) := qt\left(g(n) + r(n)b^k, s^{n+1}\right) \\ r(n+1) := rs\left(g(n) + r(n)b^k, s^{n+1}\right). \end{cases}$$

Remember that qt and rs are the quotient and rest functions, respectively. By definition, we have r (n) < s^n for all n. This implies immediately the last equality of (8). It remains only to show that h (n) < d for all n which follow from the following inequalities:

$$h(0) \cdot s \le g(0) \le b^k - 1 < b^k = d \cdot s$$

$$h(n + 1) \cdot s^{n+1} \le g(n) + r(n)b^k \le (b^k - 1) + (s^n - 1)b^k = s^n b^k - 1$$
$$= s^{n+1}d - 1 < d \cdot s^{n+1}.$$

Thus, h(n) < d for all n and x = $\sum_{n=0}^{\infty} h(n) d^{-(n+1)}$ is a p.r. d-adic expansion.

The inverse of the Theorem 5.5 was an open question in Mostowski [10]. A positive answer to this question was given by Lachlan [8].

Theorem 5.6 (Lachlan [8]): Let b, d > 1, $R_2^b \subseteq R_2^d$ if and only if d divides a power of b, i.e., $(\exists k, s) (b^k = s \cdot d)$.

To prove this theorem, Lachlan shows an equivalent characterization of the class R_2^b as follows. Let R (b):= {m $\cdot$ b^{-n}: n, m $\in \mathbb{N}$ } be the class of all b-adic rational numbers. For any set A of rational numbers, denote by C_A^0 the class of real number x such that the Dedekind cut of x restricted to A(i.e., the intersection $C_x \cap A$) is primitive recursive. Then Lachlan shows that $R_2^b = C_{R(b)}^0$, for any b > 1. That is, a real number x is b-adic p.r. if and only if the set {(n, m): m $\cdot$ b^{-n} < x & n, m $\in \mathbb{N}$ } is p.r.

Furthermore, Lachlan shows that, $C_A^0 \ne \emptyset$, if A is p.r. dense (roughly, for any rational numbers x<y, a rational number z between x and y can be found primitive recursively). For any natural number b, d > 1, if no power of b is divisible by d, then R(b)\R(d)is p.r. dense. This implies that $R_2^b \setminus R_2^d = C_{R(b)}^0 \setminus C_{R(b)}^0 \ne \emptyset$.

THE DEDEKIND CUT REPRESENTATION

This section discusses the real numbers which have p.r. Dedekind cuts. We will see that, these real numbers can be described equivalently in four different ways. By definition, the (left) Dedekind cut of a real number x is the set $C_x := \{r \in \mathbb{Q} : r < x\}$ of rational numbers. Thus, x has a p.r. Dedekind cut means that the set Cx is a p.r. set, i.e., the characteristic function $\chi_x : \mathbb{Q} \to \{0, 1\}$ is p.r. Here we need a notion of p.r. functions from rational numbers to natural numbers which can be defined by representing rational numbers as integer pairs. However, we can avoid this by considering the relation L_x defined by $L_x(m, n) \Longleftrightarrow: m/(n + 1) < x$ and say that x has a p.r. Dedekind cut if the relation Lx is p.r. The real numbers which have p.r. Dedekind cuts are called Dedekind p.r. and the class of all Dedekind p.r. reals is denoted by R_3.

Obviously, for any positive real number x and natural numbers n, m, we have $m/(n + 1) < x$ if and only if $m \leq \lfloor (n + 1).x \rfloor$ where $\lfloor y \rfloor$ denotes the integer part of the real number y, i.e., the maximal natural number t such that $t \leq y$. The function $f(n) := \lfloor n.x \rfloor$ is also called Beatty function or Beatty sequence of x after the Beatty's Theorem which asserts that the set $\{\lfloor nx \rfloor : n \in \mathbb{N}\}$ and $\{\lfloor nx \rfloor : n \in \mathbb{N}\}$ partitions natural numbers, if the positive irrational numbers x, y satisfy $1/x + 1/y = 1$ (see, e.g. [3]).

By the above observation, we have immediately the following description of Dedekind p.r. real numbers.

Theorem 6.1 (Peter [13]): A real number is Dedekind p.r. if and only if its Beatty function is p.r.

Another description of Dedekind p.r. real numbers uses the Hurwitz's characteristic of real numbers based on the Farey sequences ([4]). In mathematics, the Farey sequence of order n is the increasing sequence of irreducible fractions between 0 and 1 which have denominators less than or equal to n. For example, the Farey sequence of order four is F_4

$= \{0/1, 1/4, 1/3, 1/2, 2/3, 3/4, 1/1\}$. In general, the Farey sequence F_n of order n is an increasing sequence of irreducible fractions s/t such that $0 \le s \le t \le n$ and $t \ne 0$. One of the most interesting properties, due to Haros (see [2]), of Farey sequence is that, for any three successive terms s_1/t_1, s_2/t_2 and s_3/t_3 of F_n, the middle one is always the "mediant" of its neighbourhoods, i.e., $s_2/t_2 = (s_1 + s_3)/(t_1 + t_3)$.

By means of Farey sequence, Hurwitz ([4]) describes an irrational real number $x \in (0, 1)$ by a function $\gamma_x: \mathbb{N} \to \{0, 1\}$ which is called Hurwitz characteristic of x and is defined as follows

Initially, let $\gamma_x(0) = 0$. To define $\gamma_x(1)$, notice that x is between two elements 0/1 and 1/1 of the Farey sequence of order 1. Comparing x with the mediant (i.e., $(0 + 1)/(1 + 1)$) of these elements and define $\gamma_x(1) :=$ 0 if $x < 1/2$ and $\gamma_x(1) := 1$ if $x > 1/2$.

Suppose that $\gamma_x(n)$ is defined and x is located between two adjacent fractions in some Farey sequence of the lowest order which have been used sofar, say s_1/t_1 and s_2/t_2. Then defined $\gamma_x(n+ 1) := 0$ if $x < (s_1 + s_2)/(t_1 + t_2)$, and $\gamma_x(n + 1) := 1$ if $x > (s_1 + s_2)/(t_1 + t_2)$.

The Hurwitz characteristic supplies a simply way to find the continued fraction of a real number. We will explain in the proof of Theorem 7.4 in Section 7. The following theorem gives a new characterization of the Dedekind p.r. real numbers.

Theorem 6.2 (Lehman [9]): A real number $x \in (0, 1)$ is Dedekind p.r. if and only if its Hurwitz characteristic γ_x is p.r.

Proof: Suppose that $x \in (0, 1)$ is Dedekind p.r., i.e., L_x is p.r. Define a function e by $e(m, n) = 1$ if $x < m/(n + 1)$ and $e(m, n) = 0$ if $x > m/(n + 1)$.

Then e is p.r. Now the Hurwitz characteristic γ_x can be defined more formally with help of four additional functions s_1, s_2, t1 and t_2 as follows.

Initially let $\gamma_x(0) := 0$, $s_1(0) := 0$ and $s_2(0) = t_1(0) = t_2(0) := 1$. At the stage n, suppose that x locates between $s_1(n)/t^1(n)$ and $s_2(n)/t_2(n)$. Then define $\gamma_x(n+1) = 0$ or 1 depending on whether $x < \dfrac{s_1(n) + s_2(n)}{t_1(n) + t_2(n)}$ or not. That is, we have

$$\gamma_x(n+1) = e\,(s_1(n) + s_2(n), t_1(n) + t_2(n) - 1).$$

Now x locates between new fractions $\dfrac{s_1(n+1)}{t_1(n+1)}$ and $\dfrac{s_2(n+1)}{t_2(n+1)}$ of some Far-

ey sequence which equal to $\dfrac{s_2(n)}{t_2(n)}$ and $\dfrac{s_1(n) + s_2(n)}{t_1(n) + t_2(n)}$ if $\gamma_x(n+1) = 0$, or

$\dfrac{s_1(n) + s_2(n)}{t_1(n) + t_2(n)}$ and $\dfrac{s_2(n)}{t_2(n)}$ otherwise, respectively. That is, we have

$$
\begin{cases}
s_1(n+1) = s_1(n) + \gamma_x(n+1)s_2(n), \\
t_1(n+1) = t_1(n) + \gamma_x(n+1)t_2(n). \\
s_2(n+1) = s_2(n) + (1 - \gamma_x(n+1))s_1(n). \\
t_2(n+1) = t_2(n) + (1 - \gamma_x(n+1))t_1(n).
\end{cases}
$$

$$(10)$$

Thus, all functions γ_x, s_1, s_2, t_1, t_2 defined by equations (9) and (10) are p.r. and especially, x has a p.r. Hurwitz characteristic. On the other hand, suppose that x has a p.r. Hurwitz characteristic γ_x. Then we can define four p.r. functions s_1, s_2, t_1 and t_2 according to (10). By a simple induction, we can see that max $\{t_1(n), t_2(n)\} > n$ for all n. Con-

sequently $\dfrac{s_1(n)}{t_1(n)}$ and $\dfrac{s_2(n)}{t_2(n)}$ are adjacent fractions in some Farey series

of order greater than n. It follows that there can be no number m/n

such that $\dfrac{s_1(n)}{t_1(n)} < \dfrac{m}{n} < \dfrac{s_2(n)}{t_2(n)}$. Hence we have $\lfloor nx \rfloor = \lfloor ns_1(n)/t_1(n) \rfloor =$

$\lfloor ns_2(n)/t_2(n) \rfloor$. So x has a p.r. Dedekind cut by Theorem 6.1.

Now we discuss the relation between Dedekind p.r. and b-adic real numbers. Notice that, any finite initial segment of a b-adic expansions of a real number x corresponds to a rational number which is less (or equal, if x is rational) than x. Thus, it is possible from a p.r. Dedekind cut of x to find p.r. b-adic expansions of x. This can be done in a primitive recursive way. The other direction is impossible as shown by Specker [18].

Theorem 6.3 (Specker [18]): Let b > 1. Any Dedekind p.r. real is b-adic p.r. But there is a b-adic p.r. real which is not Dedekind p.r. That is, R_3 $\subsetneq R_2^b$.

Proof: If $x \in [0, 1]$ is Dedekind p.r., then, by Theorem 6.1, the Beatty function $\lfloor n . x \rfloor$ is a p.r. function. The b-adic expansion f of x can be defined recursively by $f(0) := \lfloor x \rfloor$ and, for any n,

$$f(n+1) := \max \left\{ t \le b : \sum_{i=0}^{n} f(i) \cdot b^{-(i+1)} + t \cdot b^{-(n+2)} \le x \right\}$$

$$= \max \left\{ t \le b : \sum_{i=0}^{n} f(i) \cdot b^{(n+1-i)} + t \le \lfloor b^{(n+2)} x \rfloor \right\}$$

Thus, x is b-adic expansion p.r. and hence $R_3 \subseteq R_2^b$

To prove the inequality $R_3 \ne R_2^b$, assume by contradiction that $R_3 = R_2^b$ for some b > 1. Choose a natural number d such that d does not divide b^k for any $k \in \mathbb{N}$. According to Theorem 5.6, we have $R_3 = R_2^b \subsetneq R_2^b$. This contradicts the fact $R_3 \subseteq R_2^b$.

By Theorem 6.3, we cannot always get a p.r. Dedekind cut of a real x from a p.r. b-adic expansion of x. However, the situation is different, if x can be represented in b-adic expansion primitive recursively and uniformly in all bases b. Here the uniform dependence of a b-adic expansion to its base b refers to the dependence of each digit to the base. This can be described by a "uniform digits function". Precisely, a function f: $\mathbb{N}^2 \to \mathbb{N}$ is called a uniform base expansion of a real number x if, for all n $\in \mathbb{N}$ and any natural number b $\geq$ 2,

$$0 \leq f(b, n) < b \text{ and } x = \sum_{n=0}^{\infty} f(b, n)b^{-(n+1)}.$$

(11)

If this function f is p.r., then we say that x has a p.r. uniform base expansion. The following theorem shows the relationship between p.r. uniform base expansions and p.r. Dedekind cuts.

Theorem 6.4: A real number has a p.r. uniform base expansion if and only if it is Dedekind p.r.

Proof: Suppose that x $\in$ [0, 1] has a p.r. uniform base expansion f_x, i.e., the p.r. function f_x satisfies the condition (11). This means that, for any natural number b > 1, we have

$$b \cdot x = f_x(b, 0) + \frac{f_x(b, 1)}{b^1} + \frac{f_x(b, 2)}{b^2} + \cdots = \sum_{i=0}^{\infty} f(b, i)b^{-i}.$$

Thus, we have $\lfloor b.x \rfloor = f_x(b, 0)$, if x is not a rational number and hence is not of the form $x = m/b^k$. That is, the Beatty function of x is p.r. By Theorem 6.1, x is a Dedekind p.r. real number. If x is rational, then x is obviously a Dedekind p.r. real number too.

On the other hand, if x is a Dedekind p.r. real number, then its Beatty function $\lfloor n.x \rfloor$ is p.r. The uniform base expansion f_x of x can be obviously defined inductively by

$$\begin{cases} f_x(0) & := \lfloor b \cdot x \rfloor \\ f_x(n+1) := \lfloor b^{n+1} \cdot x \rfloor - b \cdot \lfloor b^n \cdot x \rfloor. \end{cases}$$

And hence f_x is p.r. That is, x has a p.r. uniform base expansion.

Between Cauchy p.r. reals and Dedekind p.r. reals Specker [18] has shown the following decomposition theorem.

Theorem 6.5 (Specker [18]): Every Cauchy p.r. real number is the sum of two Dedekind p.r. real numbers.

Proof: Let x be a Cauchy p.r. real number. Suppose w.l.o.g. that x > 1. It is not difficult to see that there is a p.r. function f: $\mathbb{N} \to \{1, 2, 3\}$ such that $x = \sum_{n=0}^{\infty} f(n)2^{-n}$. Define two p.r. functions h_0, h_1 by

$$h_0(n) := \begin{cases} 0 & \text{if } \lfloor \sqrt{n} \rfloor \text{ is even,} \\ 1 & \text{if } \lfloor \sqrt{n} \rfloor \text{ is odd;} \end{cases}$$

$$h_1(n) := \begin{cases} 1 & \text{if } \lfloor \sqrt{n} \rfloor \text{ is even,} \\ 0 & \text{if } \lfloor \sqrt{n} \rfloor \text{ is odd;} \end{cases}$$

Thus, the real number x can be decomposed into to reals s and t, i.e., x = s+t, where s and t are defined by

$$s = \sum_{n=0}^{\infty} h_0(n) f(n) \cdot 2^{-n} \quad \text{and} \quad t = \sum_{n=0}^{\infty} h_1(n) f(n) \cdot 2^{-n}.$$

It remains to show that s and t are Dedekind p.r. We consider here only the real t. The proof for s is similar.

Notice that, for any n and k, if $1 \le k \le 4n + 3$, then $4n^2 + 4n + k$ is odd and hence $h_1(4n^2 + 4n + k) = 0$. Thus we have

$$t = \sum_{k=0}^{4n^2+4n} \frac{h_1(k) f(k)}{2^k} + \sum_{k=4n^2+8n+4}^{\infty} \frac{h_1(k) f(k)}{2^k}.$$

Denote two partial sums by $t_1(n)$ and $t_2(n)$, respectively, and let

$$w(n) = n \cdot 2^{4n^2+4n} \cdot t_1(n) \quad \text{and} \quad R(n) = n \cdot 2^{4n^2+4n} \cdot t_2(n).$$

Then ω is a p.r. function and $0 \leq R(n) < 1$ for all n. This implies that

$$m/n < t \iff m \cdot 2^{4n^2+4n} < \omega(n) + R(n) \iff m \cdot 2^{4n^2+4n} < \omega(n).$$

That is, t has a p.r. Dedekind cut.

By Theorem 6.5, the class R_1 of Cauchy p.r. reals is the closure of the class R_3 of Dedekind p.r. reals under arithmetical operations.

THE CONTINUED FRACTION EXPANSION REPRESENTATION

The continued fraction is another very interesting representation of real numbers. As we have mentioned in Section 2, any irrational number x can represented as an infinite continued fraction $x = [b_0, b_1, b_2, \cdots]$ where $b_n \geq 1$ for $n \in \mathbb{N}$. In this section, we use "real numbers" to refer to just "irrational numbers" for simplicity since the technical results for the cases of rational numbers are trivial and obvious.

For $x = [b_0, b_1, b_2, \cdots]$ and $n \in N$, the finite continued fraction $x_n := [b_0, b_1, b_2, \cdots , b_n]$ is a rational number and can be denoted by u_n/v_n. By simple calculation, u_n, v_n can be determined inductively as follows

$$\begin{cases} u_{-1} = 1 \quad v_{-1} = -1, \quad u_0 = b_0, \quad v_0 = 1 \\ u_{n+2} = b_{n+2}u_{n+1} + u_n, \\ v_{n+2} = b_{n+2}v_{n+1} + v_n. \end{cases}$$

$$(12)$$

Here the terms u_{-1}, v_{-1} are defined for the technical simplicity. The fractions u_n/v_n are called the convergent of order n and they are reduced fractions for all n. By a simple induction we can show that $v_n < v_{n+1}$ for all $n \geq 1$ and hence $v_n \geq n$ for all n. Each convergent is nearer to x than the preceding convergent. In addition, the convergents provide the best approximations to x in the following sense: if $n > 1$, $0 < v \leq v_n$, and $u/v \neq u_n/v_n$, then $|x - u_n/v_n| < |x - u/v|$.

About the convergents of a continued fraction expansion, we have the following further properties which will be used in the proofs later on.

Lemma 7.1 (cf. [2, 12]): Let $x = [b_0, b_1, b_2, \cdots]$ and let u_n/v_n be its convergent of order n. Then we have

(i) $u_n v_{n-1} - v_n u_{n-1} = (-1)^{n-1};$

(ii) $b_{n+1} = \left\lfloor \frac{u_{n-1} - x v_{n-1}}{x v_n - u_n} \right\rfloor;$

(iii) $\frac{1}{v_n(v_{n+1}+v_n)} < \left| x - \frac{u_n}{v_n} \right| < \frac{1}{v_n v_{n+1}};$

(iv) $\frac{u_0}{v_0} < \frac{u_2}{v_2} < \frac{u_4}{v_4} < \cdots < x < \cdots < \frac{u_5}{v_5} < \frac{u_3}{v_3} < \frac{u_1}{v_1}.$

Now we are going to investigate the class of real numbers which have a p.r. continued fractions. By (12), if x has a p.r. continued fraction, then the corresponding sequences (u_s) and (v_s) are p.r. too. That is, the sequence (u_s/v_s) is a p.r. approximation of x. According to Lemma 7.1.(iii), it is easy to see that x is Cauchy p.r. Of course, we can do better. From items 3 and 4 of the Lemma 7.1 and the fact that $v_n \geq n$, we have

$$0 < nx - \frac{n u_{2n}}{v_{2n}} < \frac{n}{v_{2n} v_{2n+1}} \leq \frac{1}{v_{2n}},$$

and hence

$$\frac{n u_{2n}}{v_{2n}} < nx < \frac{n u_{2n}}{v_{2n}} + \frac{1}{v_{2n}}.$$

In order to change the integer part of $\frac{n u_{2n}}{v_{2n}}$, we must add at least $\frac{1}{v_{2n}}$ to it. This implies that

$$\left\lfloor \frac{nu_{2n}}{v_{2n}} \right\rfloor \leq \lfloor nx \rfloor < \left\lfloor \frac{nu_{2n}+1}{v_{2n}} \right\rfloor \leq \left\lfloor \frac{nu_{2n}}{v_{2n}} \right\rfloor + 1.$$

That is, the Beatty function $\lfloor nx \rfloor = \lfloor nu_{2n} / v_{2n} \rfloor$ of x is primitive recursive. By Theorem 6.1, x is Dedekind p.r., i.e., $R_4 \subseteq R_3$. This result belongs to Lehman [9]. The next natural question is, whether there exists a Dedekind real which does not have a p.r. continued fraction? To answer this question, Lehman shows another characterization of the real numbers which have p.r. continued fraction by means of primitive-recursively irrationality of P´eter [13,14].

Roughly speaking, an irrational number x is called primitive-recursively irrational (p.r. irrational for short) if it is possible to find a primitive-recursively lower bound of the distance between x and any given rational number. More precisely, there is a p.r. function f such that for all positive integers m and n

$$\left| x - \frac{m}{n} \right| > \frac{1}{f(n)}.$$

(13)

P´eter [13, 14] used a slightly different but equivalent definition and she used the name recursively irrational because in P´eter [13] recursive means actually primitive recursive. The name primitive-recursively irrational was used by Goodstein [1] where it is shown that π is p.r. irrational.

P´eter [13] shows that a Cauchy p.r. real is continued fraction p.r. if it is p.r. irrational. Lehman [9] shows that this is in fact a necessary and sufficient condition of a real number with p.r. continued fraction.

Theorem 7.2 (Lehman [9]): A real number x has a p.r. continued fraction expansion if and only if it is Cauchy p.r. and p.r. irrational.

Proof: Suppose that x = [b_0, b_1, b_2, ···] is a p.r. continued fraction. According to (12), the sequence (u_n/v_n) of the convergents of x is obviously p.r. From Lemma 7.1, it is easy to see that x is Cauchy p.r. To show that x is p.r. irrational, it suffices to look at the following inequality.

$$\left| x - \frac{m}{n} \right| \geq \left| x - \frac{u_n}{v_n} \right| > \frac{1}{v_n(v_n + v_{n+1})}$$

For m, n > 0. Here the first inequality follows from the fact that (u_n/v_n) is the best approximation to x and $n \leq v_n$ and the second inequality follow from Lemma 7.1. (iii). Thus, the p.r. function $f(n) := v_n(v_n + v_{n+1})$ witnesses that x is p.r. irrational.

For the other direction, suppose that x is a p.r. irrational Cauchy p.r. real number. Then there are p.r. functions c, d, f such that

$$\left| x - \frac{c(n)}{d(n)} \right| \leq \frac{1}{n} \text{ and } \left| x - \frac{m}{n} \right| > \frac{1}{f(n)}$$

(14)

For all positive integers m, n. Assume w.l.o.g. that the function f is monotone increasing. Otherwise we can consider the function $f'(n) :=$ max {f (m): m ≤ n} instead.

Let $x = [b_0, b_1, b_2, \cdots]$ and u_n/v_n be its convergent of order n. We want to show that (b_n) is a p.r. sequence. Define a p.r. function $h(n) := f^{(n+1)}(v_1)$, i.e., $h(0) := f(v_1)$ and $h(n + 1) := fh(n)$. Then, by item (iii) of Lemma 7.1 and the second part of (14), we have $1/f(v_n) < |x - u_n/v_n| < 1/v_n v_{n+1} \leq 1/v_{n+1}$. That is, we have $v_{n+1} \leq f(v_n)$ for all n. By a simple induction we can show that $f(v_{n+1}) \leq h(n)$ which implies immediately that

$$\left| x - \frac{u_{n+1}}{v_{n+1}} \right| > \frac{1}{f(v_{n+1})} \geq \frac{1}{h(n)} \geq \left| x - \frac{ch(n)}{dh(n)} \right| = |x - x_n|$$

(15)

Where $x_n := \dfrac{ch(n)}{dh(n)}$. Since x is between $\dfrac{u_n}{v_n}$ and $\dfrac{u_{n+1}}{v_{n+1}}$ and it is closer to

$\dfrac{u_{n+1}}{v_{n+1}}$ than to $\dfrac{u_n}{v_n}$. So x_n must lie between $\dfrac{u_n}{v_n}$ and $\dfrac{u_{n+1}}{v_{n+1}}$ too. Since the rational numbers un vn and un+1 vn+1 have the same partial quotients of order less than n+ 1 which are partial quotients of order less than n+ 1 of the real x and the rational number x_n too (see [12], p. 35). Thus, it suffices to show that the sequence (b_n) can be defined primitive-recursively from the p.r. sequence (x_n). By item (iii) of the Lemma 7.1, this can be realized by the following definition combining with the equations (12).

$$\begin{cases} b_0 = \lfloor x_0 \rfloor \\ b_{n+1} = \left\lfloor \dfrac{u_{n-1} - x_{n+1} v_{n-1}}{x_{n+1} v_n - u_n} \right\rfloor \end{cases} \tag{16}$$

Therefore (b_n) is a p.r. sequence and hence x has a p.r. continued fraction expansion.

By means of the characterization of the Theorem 7.2, Lehman [9] can further show not every Dedekind p.r. real has a p.r. continued fraction expansion.

To this end, we need another technical lemma which can be easily proved from the Lemma 5.1.

Lemma 7.3 (Lehman [9]): There is a p.r. function $\lambda: \mathbb{N} \to \{1, 2,\}$ which takes value 1 infinitely many times such that the function σ defined by $\sigma(n) := \mu m \geq n \ (\lambda(m) \neq 2)$ is not p.r.

Theorem 7.4 (Lehman [9]): There is a Dedekind p.r. real number x which is not primitive recursively irrational.

Proof: The real number x is given by its Hurwitz characteristic γ_x which is defined by $\gamma_x(0) = 0$ and

$$\gamma_x(n+1) := \begin{cases} \gamma_x(n) & \text{if } \lambda(n) = 2, \\ 1 \div \gamma_x(n) & \text{if } \lambda(n) \neq 2. \end{cases}$$

Where λ is the p.r. function of Lemma 7.3. By Theorem 6.2 x is Dedekind p.r.

As it is shown by Hurwitz [4], the continued fraction x = [0, b_1, b_2, b_3, ...] of x can be easily obtained from the Hurwitz characteristic γ_x of x by simply counting the successive 0's and 1's. Namely, b_1 is the number of leading 0's of the sequence $(\gamma_x(n))$. Then, the next b2 values are 1, the next b3 values are 0, etc., where $b_i \geq 1$. That is, the sequence $(\gamma_x(n))$ have the following form.

$$\underbrace{0\ 0\ \cdots\ 0}_{b_1}\ \underbrace{1\ 1\ \cdots\ 1}_{b_2}\ \underbrace{0\ 0\ \cdots\ 0}_{b_3}\ \underbrace{1\ 1\ \cdots\ 1}_{b_4}\ 0\cdots\cdots$$

By the definition of γx, we have γx(n) $\neq$ γx(n + 1) if and only if λ(n) $\neq$ 2. Thus, for any n, the first natural number m after n such that λ(m) $\neq$ 2 is bounded above by $h(n) := \sum_{i \leq n+1} b_i$. If (b_n) is p.r., then h is a p.r. function too. In this case, the function σ of the Lemma 7.3 should be p.r. because

$$\sigma(n) = (\mu m \geq n)(\lambda(m) \neq 2)$$
$$= (\mu m \leq h(n))(m \geq n\ \&\ \lambda(m) \neq 2).$$

This contradicts Lemma 7.3. Hence we conclude that (b_n) is not p.r. and x does not have a p.r. continued fraction expansion. By Theorem 7.2, x is nor p.r. irrational.

Corollary 7.5 (Lehman [9]): The class of real numbers which have p.r. continued fraction expansions is a proper subset if the class of Dedekind p.r. real numbers, i.e., $R_4 \subsetneq R_3$.

CONCLUSIONS

In this paper we summarize several known results about primitive recursiveness of real numbers under different representations which are scattered in literatures and show some new relations among them as well. We have seen that, the p.r. reals under different representations form a comprehensive hierarchy:

$$\mathbf{CF}(\mathcal{F}) \subsetneq \mathbf{DC}(\mathcal{F}) \subsetneq \mathbf{bAE}(\mathcal{F}) \subsetneq \mathbf{CS}(\mathcal{F}) \subsetneq \mathbf{NI}(\mathcal{F}) = \mathbf{EC}.$$

For the class $\mathcal{F}$ of p.r. functions. Among these classes, it seems that the class CS ($\mathcal{F}$) might be properly regarded as the class of primitive recursive reals. And we also see that there is a hierarchy inside bAE($\mathcal{F}$) = R_2^b for different b's in the primitive recursive level. Their relation is that R_2^b $\subseteq R_2^b$ if and only if d divides a power of b.

It is also very natural to discuss these classes for other function classes F. For example, Ko [6,7] have shown a similar hierarchy if F is the class of all polynomial time computable functions.

REFERENCES

1. Goodstein, Reuben Louis. The recursive irrationality of π. Journal of Symbolic Logic, 19 (1954), 267–274.
2. Hardy, G. H. and E. M. Wright. "An introduction to the theory of numbers". The Clarendon Press, Oxford University Press, New York, 1979.
3. Honsberger, Ross. "Ingenuity in mathematics". 6th printing. Mathematical Association of America, New Mathematical Library. 23. Washington, DC. 1998.
4. Hurwitz, A. Uber die angen¨ ¨ aherte darstellung der zahlen durch rationale br¨uche. Mathematische Annalen, 44 (1894), 417–436.
5. Kleene, Stephen Cole. "Introduction to metamathematics". D. Van Nostrand Co., Inc., New York, 1952.
6. Ko, Ker-I. On the definitions of some complexity classes of real numbers. Math. Systems Theory, 16 (1983), 95–109.
7. Ko, Ker-I. "Complexity Theory of Real Functions". Progress in Theoretical Computer Science. Birkh¨auser, Boston, MA, 1991.

8. Lachlan, A.H. Recursive real numbers. Journal of Symbolic Logic, 28 (1963), 1–16.
9. Lehman, R.S. On primitive recursive real numbers. Fundamenta Mathematicae, 49 (1960/61), 105–118.
10. Mostowski, Andrzej. On computable sequences. Fundamenta Mathematicae, 44 (1957), 37–51.
11. Myhill, John. Criteria of constructibility for real numbers. Journal of Symbolic Logic, 18 (1953), 7–10.
12. Perron, Oskar. "Die Lehre von den Kettenbr¨uchen. Dritte, verbesserte und erweiterte Aufl. Bd. II. Analytisch-funktionentheoretische Kettenbr¨uche". B. G. Teubner Verlagsgesellschaft, Stuttgart, 1957.
13. P´eter, R´ozsa. Zum begriff der rekursiven reellen zahlen. Acta Scientiarum Mathematicarum Szeged, 12/A (1950), 239–245.
14. P´eter, R´ozsa. "Rekursive Funktionen". Akademischer Verlag, Budapest, 1951.
15. Rice, H. Gordon. Recursive real numbers. Proc. Amer. Math. Soc, 5 (1954), 784–791.
16. Robinson, Raphael M. Review of "Peter, R., Rekursive Funktionen". Journal of Symbolic Logic, 16 (1951), 280–282.
17. Skordev, Dimiter. Characterization of the computable real numbers by means of primitive recursive functions. In Jens Blanck, Vasco Brattka, and Peter Hertling, editors, Computability and Complexity in Analysis (4th International Workshop, CCA 2000, Swansea, UK, September 2000), volume 2064 of Lecture Notes in Computer Science, pages 296–309, Springer, Berlin, 2001.
18. Specker, Ernst. Nicht konstruktiv beweisbare S¨atze der Analysis. Journal of Symbolic Logic, 14 (1949), 145–158.
19. Turing, Alan M. On computable numbers, with an application to the "Entscheidungsproblem". Proceedings of the London Mathematical Society, 42 (1936), 230–265.
20. Weihrauch, Klaus. "Computable Analysis, An Introduction". Springer, Berlin Heidelberg, 2000.

CITATION

Qingliang Chen, Kaile Su, Xizhong Zheng, Primitive Recursiveness of Real Numbers under Different Representations, Electronic Notes in Theoretical Computer Science, Volume 167, 24 January 2007, Pages 303-324, ISSN 1571-0661, doi. org/10.1016/j.entcs.2006.08.018.

Comparison of Volumes of Convex Bodies in Real, Complex, and Quaternionic Spaces

Boris Rubin
Department of Mathematics, Louisiana State University, Baton Rouge, LA 70803, USA

ABSTRACT

The classical Busemann–Petty problem (1956) asks, whether origin-symmetric convex bodies in Rn with smaller hyperplane central sections necessarily have smaller volumes. It is known, that the answer is affirmative if n ≤ 4 and negative if n > 4. The same question can be asked when volumes of hyperplane sections are replaced by other comparison functions having geometric meaning. We give unified analysis of this circle of problems in real, complex, and quaternionic n-dimensional spaces. All cases are treated simultaneously. In particular, we show that the Busemann–Petty problem in the quaternionic n-dimensional space has an affirmative answer if and only if n = 2. The method relies on the properties of cosine transforms on the unit sphere. We discuss possible generalizations.

INTRODUCTION

Real and complex affine and Euclidean spaces are traditional objects in integral geometry. Similar spaces can be built over more general algebras, in particular, over quaternions. The discovery of quaternions is attributed to W.R. Hamilton (1843).[1] A variety of problems of differential geometry in quaternionic and more general spaces over algebras were investigated by Rosenfel'd and his collaborators, in particular, in the Kasan' geometric school (Russia); see, e.g., [49,69,65]. Some problems of quaternionic integral geometry, mainly related to polytopes and invariant densities, were studied by Coxeter, Cuypers, Santaló, and others; see [11,24,23,61] and references therein. One should also mention a series of papers by Markina and her collaborators, related to geometric analysis on division algebras; see, e.g., [10].

In the present article we are focused on comparison problems for convex bodies in the general context of the space $_n$, where stands for the field of real numbers, the field $\mathbb{C}$ of complex numbers, and the skew field $\mathbb{H}$ of real quaternions. Since $\mathbb{H}$ is not commutative, special consideration is needed in this case.

Let, for instance, K and L be origin-symmetric convex bodies in $_n$ with section functions

$$S_K(H) = \mathrm{vol}_{n-1}(K \cap H) \quad \text{and} \quad S_L(H) = \mathrm{vol}_{n-1}(L \cap H)$$

H being a hyperplane passing through the origin. Suppose that $S_K(H) \leq$ SL(H) for all such H. Does it follow that $\mathrm{vol}_n(K) \leq \mathrm{vol}_n(L)$? Since the latter may not be true, another question arises: For which operator D is the implication valid?

$$\left(\mathcal{D}S_K(H) \leq \mathcal{D}S_L(H), \ \forall H \right) \implies \mathrm{vol}_n(K) \leq \mathrm{vol}_n(L)$$

Comparison problems of this kind attract considerable attention in the last decade, in particular, thanks to remarkable connections with harmonic analysis. The first question is known as the Busemann–Petty (BP) problem [9]. Many authors contributed to its solution, e.g., Ball [3], Barthe, Fradelizi and Maurey [4], Gardner [13,14,16], Giannopoulos [18], Grinberg and Rivin [21], Hadwiger [25], Koldobsky [33], Larman and Rogers [38], Lutwak [40], Papadimitrakis [44], Rubin [54], Zhang [73]. The answer is really striking. It is "Yes" if and only if n≤4; see [15,16,33,35], and references therein.

The second question, related to implication (1.1), was asked by Koldob-sky, Yaskin and Yaskina [36]. It was called the modified Busemann–Petty problem. Both questions were studied by Koldobsky, König and Zymonopoulou [34] and Zymonopoulou [74] for convex bodies in. $\mathbb{C}^n$.

The answer to the first question for $\mathbb{C}^n$ is "Yes" if and only if $n \leq 3$.

We suggest a unified analysis of these problems for real, complex, and also quaternionic n-dimensional spaces and the relevant $(n - 1)$-dimensional subspaces H. All these cases are treated simultaneously. In particular, we show that the quaternionic BP problem has an affirmative answer if and only if $n = 2$.

The article is nearly self-contained. Our proofs essentially differ from those in the aforementioned publications and rely on the properties of the generalized cosine transforms on the unit sphere [51,52,56].

The setting of the quaternionic BP problem and its solution require careful preparation and new geometric concepts. The crux is that, unlike the fields of real and complex numbers, the algebra of quaternions is not commutative. This results in non-uniqueness of quaternionic analogues of such concepts as a vector space and its subspaces, a symmetric convex body, a norm, etc.

Another motivation for our work is the lower dimensional Busemann–Petty problem (LDBP), which sounds like the usual BP problem, but the hyperplane sections are replaced by plane sections of fixed dimension $1 < i < n-1$, In the case $i=2$, $n=4$ an affirmative answer to LDBP follows from the solution of the usual BP problem For $i > 3$, a negative answer was first given a negative answer was first given by Bourgain and Zhang [8]; see also [33,59] for alternative proofs. In the cases $i = 2$ and $i = 3$ for $n > 4$, the answer is generally unknown, however, if the body with smaller sections is a body of revolution, the answer is affirmative; see [22,72,59]. The paper [57] contains a solution of the LDBP problem in the more general situation, when the body with smaller sections is invariant under rotations, preserving mutually orthogonal subspaces of dimensions ℓ and $n-\ell$, respectively. The answer essentially depends on ℓ.

It is natural to ask, how invariance properties of bodies affect the corresponding LDBP problem.

Of course, this question is too vague, however, every specific example might be of interest. The article [34] on the BP problem in $\mathbb{C}^n$ actually deals with the LDBP problem for $(2n - 2)$- dimensional sections of 2n-dimensional convex bodies, which are invariant under the block scalar subgroup G of SO(2n) of the form

$$G = \left\{ g = \text{diag}(g_1, \ldots, g_n) : g_1 = \cdots = g_n \in SO(2) \right\}.$$

The latter means that the corresponding bodies in $\mathbb{C}^n$ are invariant under the transformation for every $\theta \in \mathbb{R}$.

$$\mathbb{C}^n \ni x \rightarrow e^{\sqrt{-1}\,\theta} x \in \mathbb{C}^n$$

This symmetry of convex bodies is often called "circular" in the complex variables literature.

We will show that the BP problem in the n-dimensional left and right quaternionic spaces $\mathbb{H}_l^n$ and $\mathbb{H}_r^n$ is equivalent to the LDBP problem for $(4n - 4)$-dimensional sections of 4n-dimensional convex bodies, which are invariant under a certain subgroup $G \subset SO(4n)$ of block diagonal matrices, having n equal 4×4 isoclinic (or Clifford) blocks. Every such block is a left (or right) matrix representation of a real quaternion and has the property of rotating all lines through the origin in $\mathbb{R}^4$ by the same angle. We give complete solution to this "G-invariant" comparison problem in the general contest of dn-dimensional convex bodies, $n > 1$, the symmetry of which is determined by complete system of orthonormal tangent vector fields on the unit sphere S^{d-1}. A classical result of differential topology says that such systems are available only on S^1, S^3, and S^7; see Section 2.6. We also study the corresponding modified BP, when the "derivatives" DS_K and DS_L are compared.

Plan of the Paper and Main Results

The most significant part of the paper (Sections 2–4) deals with necessary preparations, the aim of which is to make the text accessible to a broad audience of analysts and geometers. Section 2.1 contains basic notation. In Sections 2.2 and 2.3 we recall basic facts about quaternions and vector spaces $\mathbb{K}^n$, $\mathbb{K} \in \{,,\mathbb{H}\}$. This information is scattered in the literature; see, e.g., [37,39,46,45,67,70,71]. We present it in the form, which is suitable for our purposes. Since $\mathbb{H}$ is not commutative, we have to distinguish the left vector space $\mathbb{H}_l^n$ and the right vector space $\mathbb{H}_r^n$.

In Section 2.4 we introduce a concept of equilibrated body in the general context of the space $\mathfrak{A}^n$, where $\mathfrak{A}^n$ is a real associative normed algebra. These bodies serve as a substitute for the class of origin-symmetric convex bodies in $\mathbb{R}^n$. As in the real case (see, e.g., [5]), they are associated with norms on $\mathfrak{A}^n$. In the complex case, other names ("circular", "absolutely convex", or "balanced") are also in use [19,28,48]. We could not find any description of this class of bodies in the quaternionic or more general contexts and present this topic in detail.

In Section 2.5 we give precise setting of the hyperplane slice comparison problem of the Busemann–Petty type for equilibrated convex bodies in $\mathbb{K}^n \in \{\mathbb{R}^n, \mathbb{C}^n, \mathbb{H}_l^n \; \mathbb{H}_r^n\}$ (see Problem A), the corresponding lower dimensional slice comparison problem for G-invariant convex bodies in $\mathbb{R}^N$, $N = dn$ (Problem B), and also the more general Problem C, where comparison of slices is realized in terms of certain derivatives of slicing functions. Here $d = 1, 2$, and 4, which corresponds to the real, complex, and quaternionic cases, respectively. Section 2.6 contains necessary information about vector fields on spheres and extends Problems B and C to the octonionic case $d = 8$. This value of d cannot be increased in the framework of our method.

Section 3 provides the reader with necessary background from harmonic analysis related to analytic families of cosine transforms and intersection bodies. The latter were introduced by Lutwak [40] and generalized in different directions; see, e.g., Gardner [15], Goodey, Lutwak and Weil [20], Koldobsky [33], Milman [41], Rubin and Zhang [59], Zhang [72]. Here we follow our previous papers [51,52,56]. We draw attention to Section 3.2 devoted to homogeneous distributions and Riesz fractional derivatives $D^\alpha = (-)^{\alpha/2}$, where $\triangle$ is the Laplace operator on $\mathbb{R}^n$. An important feature of these operators is that the corresponding Fourier multiplier $|y|^\alpha$ does not preserve the Schwartz space $S(\mathbb{R}^N)$ and the phrases like "in the sense of distributions" (cf. [36,35,74]) require careful explanation and justification.

Section 4 is devoted to weighted section functions of origin-symmetric convex bodies. If K is such a body, these functions are defined as i-plane Radon transforms of the characteristic function $_{xK}(x)$ (i.e. $_{xK}(x) = 1$ when $x \in K$, and 0 otherwise), with integration against the weighted Lebesgue measure with a power weight $|x|^\beta$. The usefulness of such functions was first noted in [53] and mentioned in [59, p. 492]. Smoothness properties of these functions play a decisive role in establishing main results, and we study them in detail. Similar properties in the context of the modified BP problem in $\mathbb{R}^n$ and $\mathbb{C}^n$ were briefly indicated in [36,35,74], however, the details (which are important and fairly nontrivial) were omitted.

In Section 5 we obtain main results; see Theorems 5.4, 5.5, 5.8, and Corollaries 5.6, 5.7. In particular, the hyperplane slice comparison problem in $\mathbb{K}^n$ has an affirmative answer if and only if $n \leq 2 + 2/d$, where d = 1, 2, and 4 in the real, complex, and quaternionic cases, respectively.

PRELIMINARIES

Notation

We denote by $\sigma_{n-1} = 2\pi^{n/2}/\Gamma(n/2)$ the area of the unit sphere S^{n-1} in $\mathbb{R}^n$; SO(n) is the special orthogonal group. For $\theta \in S^{n-1}$ and $\gamma \in SO(n)$, $d\theta$ and $d\gamma$ denote the relevant probability measures; $D(S^{n-1})$ is the space of C$^\infty$-functions on Sn−1 with standard topology; De(Sn−1) is the sub-space of even functions in $D(S^{n-1})$.

In the following $M_{n,k}(\mathbb{R})$ is the set of real matrices having n rows and k columns; $M_n(\mathbb{R}) = M_{n,n}(\mathbb{R})$; A^T denotes the transpose of a matrix A; I_n $\in M_n(\mathbb{R})$ is the identity matrix; $V_{n,k} = \{F \in M_{n,k}(\mathbb{R}): FT\ F = Ik\}$ is the Sti-efel manifold of orthonormal k-frames in $\mathbb{R}^n$; $Gr_k(V)$ is the Grassmann manifold of k-dimensional linear subspaces of the vector space V .

Given a certain class X of functions or bodies and a group G, acting in the ambient space, we denote by X^G the corresponding subclass of G-invariant objects. For example, C^G(S^{n-1}) and D^G(S^{n-1}) are the spaces of continuous and infinitely differentiable functions on S^{n-1}, respectively, such that f (gθ) = f (θ), $\forall g \in G$, θ $\in$ Sn−1. The group G will be speci-fied in due course. An origin-symmetric (o.s.) star body in $\mathbb{R}^n$, n≥2, is a compact set K with non-empty interior, such that tK $\subset$ K, $\forall t \in [0, 1]$, K = −K, and the radial function ρK (θ) = sup{λ 0: λθ $\in$ K} is continuous on S^{n-1}. We denote by Kn the set of all o.s. star bodies in$\mathbb{R}^n$. A body K $\in$ K^n is said to be smooth if ρK $\in D_e(S^{n-1})$.

Quaternions

We regard $\mathbb{H}$ as a normed algebra over $\mathbb{R}$ generated by the units e0, e_1, e_2, e_3. The more familiar notation is 1, i, j, k, but we reserve these symbols for other purposes. Every element q $\in$ H is expressed as q = $q_0e_0 + q_1e_1 + q_2e_2 + q_3e_3$ (qi $\in$ R). We set

$$\bar{q} = q_0e_0 - q_1e_1 - q_2e_2 - q_3e_3, \qquad |q| = \sqrt{q_0^2 + q_1^2 + q_2^2 + q_3^2}$$

The multiplicative structure in H is governed by the rules

$$e_0e_i = e_ie_0 = e_i, \quad i = 0, 1, 2, 3,$$

$$e_1e_2 = -e_2e_1 = e_3, \qquad e_2e_3 = -e_3e_2 = e_1, \qquad e_3e_1 = -e_1e_3 = e_2,$$

$$e_1^2 = e_2^2 = e_3^2 = -e_0.$$

The product of two quaternions $p = p_0e_0 + p_1e_1 + p_2e_2 + p_3e_3$ and $q = q_0e_0 + q_1e_1 + q_2e_2 + q_3e_3$ is computed accordingly as

$$pq = (p_0q_0 - p_1q_1 - p_2q_2 - p_3q_3)e_0$$
$$+ (p_0q_1 + p_1q_0 + p_2q_3 - p_3q_2)e_1$$
$$+ (p_0q_2 - p_1q_3 + p_2q_0 + p_3q_1)e_2$$
$$+ (p_0q_3 + p_1q_2 - p_2q_1 + p_3q_0)e_3$$

$$(2.1)$$

so that

$$q\bar{q} = \bar{q}q = |q|^2, \qquad \overline{pq} = \bar{q}\bar{p}, \qquad |pq| = |p||q|, \qquad q^{-1} = \bar{q}/|q|^2 \qquad .$$

We identify

$$\mathbb{R} = \{q \in \mathbb{H}: q_1 = q_2 = q_3 = 0\}, \qquad \mathbb{C} = \{q \in \mathbb{H}: q_2 = q_3 = 0\}$$
,

and denote by $Sp(1)$ the symplectic unitary group, which is formed by quaternions of absolute value 1. There is a canonical bijection h: $\mathbb{H} \to \mathbb{R}^4$, according to which,

$$q = q_0e_0 + q_1e_1 + q_2e_2 + q_3e_3 \xrightarrow{h} v_q = (q_0, q_1, q_2, q_3)^T$$
$$Sp(1) \xrightarrow{h} S^3.$$

$$(2.2)$$

By (2.1),

$$p\bar{q} = (p_0q_0 + p_1q_1 + p_2q_2 + p_3q_3)e_0 + (-p_0q_1 + p_1q_0 - p_2q_3 + p_3q_2)e_1$$
$$+ (-p_0q_2 + p_1q_3 + p_2q_0 - p_3q_1)e_2 + (-p_0q_3 - p_1q_2 + p_2q_1 + p_3q_0)e_3$$

or

$$p\bar{q} = \sum_{i=0}^{3} (v_p \cdot A_i v_q) e_i$$

(2.3)

where v_p and v_q are defined according to (2.2),

$$A_0 = I_4 = \begin{bmatrix} 1 & 0 & 0 & 0 \\ 0 & 1 & 0 & 0 \\ 0 & 0 & 1 & 0 \\ 0 & 0 & 0 & 1 \end{bmatrix}, \quad A_1 = \begin{bmatrix} 0 & -1 & 0 & 0 \\ 1 & 0 & 0 & 0 \\ 0 & 0 & 0 & -1 \\ 0 & 0 & 1 & 0 \end{bmatrix}$$

$$A_2 = \begin{bmatrix} 0 & 0 & -1 & 0 \\ 0 & 0 & 0 & 1 \\ 1 & 0 & 0 & 0 \\ 0 & -1 & 0 & 0 \end{bmatrix}, \quad A_3 = \begin{bmatrix} 0 & 0 & 0 & -1 \\ 0 & 0 & -1 & 0 \\ 0 & 1 & 0 & 0 \\ 1 & 0 & 0 & 0 \end{bmatrix}$$

(2.4)

Similarly,

$$\bar{p}q = \sum_{i=0}^{3} (v_p \cdot A_i' v_q) e_i$$

(2.5)

$$A_0' = A_0 = I_4, \quad A_1' = \begin{bmatrix} 0 & 1 & 0 & 0 \\ -1 & 0 & 0 & 0 \\ 0 & 0 & 0 & -1 \\ 0 & 0 & 1 & 0 \end{bmatrix},$$

$$A_2' = \begin{bmatrix} 0 & 0 & 1 & 0 \\ 0 & 0 & 0 & 1 \\ -1 & 0 & 0 & 0 \\ 0 & -1 & 0 & 0 \end{bmatrix}, \quad A_3' = \begin{bmatrix} 0 & 0 & 0 & 1 \\ 0 & 0 & -1 & 0 \\ 0 & 1 & 0 & 0 \\ -1 & 0 & 0 & 0 \end{bmatrix}$$

(2.6)

One can readily see that $A_i, A_i' \in SO(4)$ and collections form orthonormal bases of $\mathbb{R}^4$ for every $q \in Sp(1)$. This gives the following.

$$\{A_0 v_q, A_1 v_q, A_2 v_q, A_3 v_q\}, \quad \{A_0' v_q, A_1' v_q, A_2' v_q, A_3' v_q\}$$

Theorem

There exist "left rotations" A_i and "right rotations" A'_i ($i = 1, 2, 3$), such that for every $\sigma \in S^3$, the frames form orthonormal bases of $\mathbb{R}^4$.

$$\{\sigma, A_1\sigma, A_2\sigma, A_3\sigma\}, \quad \{\sigma, A'_1\sigma, A'_2\sigma, A'_3\sigma\}$$

The left- and right-multiplication mappings $p \to qp$ and $p \to pq$ in $\mathbb{H}$ can be realized as linear transformations of $\mathbb{R}^4$, namely,

$$v_{qp} = L_q v_p, \qquad v_{pq} = R_q v_p \tag{2.7}$$

$$L_q = \begin{bmatrix} q_0 & -q_1 & -q_2 & -q_3 \\ q_1 & q_0 & -q_3 & q_2 \\ q_2 & q_3 & q_0 & -q_1 \\ q_3 & -q_2 & q_1 & q_0 \end{bmatrix}, \qquad R_q = \begin{bmatrix} q_0 & -q_1 & -q_2 & -q_3 \\ q_1 & q_0 & q_3 & -q_2 \\ q_2 & -q_3 & q_0 & q_1 \\ q_3 & q_2 & -q_1 & q_0 \end{bmatrix} \tag{2.8}$$

These formulas define regular representations of $\mathbb{H}$ in the algebra $M_4(\mathbb{R})$ of 4×4 real matrices:

$$\rho_l : q \to L_q, \qquad \rho_r : q \to R_{\bar{q}} \tag{2.9}$$

so that $\rho l(pq) = \rho l(p)\rho l(q)$, $\rho r(pq) = \rho r(p)\rho r(q)$. Clearly,

$$L_q = \sum_{i=0}^{3} q_i A_i, \qquad R_{\bar{q}} = \sum_{i=0}^{3} q_i A'_i \tag{2.10}$$

In particular,

$$A_i = L_{e_i}, \qquad A'_i = R_{\bar{e}_i}, \qquad i = 0, 1, 2, 3 \tag{2.11}$$

For any p,q $\in \mathbb{H}$, matrices L_p and R_q commute, that is,

$$L_p R_q = R_q L_p \tag{2.11}$$

Moreover, $\det(L_q) = \det(R_q) = |q|^4$ (see, e.g., [6, p. 28]). Since the columns of each of these matrices are mutually orthogonal, then, for $|q|$ = 1, both matrices belong to SO(4). The map is a group surjection with kernel $\{(e_0, e_0), (-e_0, -e_0)\}$ [46,70].

$$Sp(1) \times Sp(1) \to SO(4), \qquad (p, q) \to L_p R_{\bar{q}}$$

A direct computation shows that $x . R_q x = x . L_q x = q0$ for every $x \in S^3$. It means that both L_q and R_q have the property of rotating all half-lines originating from O through the same angle $\cos^{-1} q_0$ (such rotations are called isoclinic or Clifford translations [70]). We call L_q and R_q the left rotation and the right rotation, respectively. Note also that

$$J L_q J = R_{\bar{q}}, \qquad J R_q J = L_{\bar{q}}, \qquad J = \begin{bmatrix} -1 & 0 \\ 0 & I_3 \end{bmatrix} \tag{2.13}$$

This means that the left rotation becomes the right one if we change the direction of the first coordinate axis in $\mathbb{R}^4$.

Similarly, if $\mathbb{K} = \mathbb{C}$, we set

$$\mathbb{C} \ni c = a + ib \xrightarrow{h} v_c = (a, b)^T \in \mathbb{R}^2$$

so that

$$v_{cd} = v_{dc} = M_c v_d, \quad c,d \in \mathbb{C}, \quad M_c = \begin{bmatrix} a & -b \\ b & a \end{bmatrix}$$

$$(2.14)$$

Clearly, $M_c \in SO(2)$ if $|c| = 1$, and, conversely, every element of $SO(2)$ has the form M_c, $c = \cos\varphi + i\sin\varphi$.

The space $\mathbb{K}^n$

Let $\mathbb{K} \in \{\mathbb{R}, \mathbb{C}, \mathbb{H}\}$. Consider the set of "points" $x = (x_1,...,x_n)$, $xi \in \mathbb{K}$, that can be regarded as an additive abelian group in a usual way. We want to equip this set with the structure of the inner product vector space over $\mathbb{K}$. The resulting space will be denoted by $\mathbb{K}^n$. Unlike the cases $\mathbb{K} = \mathbb{R}$ and $\mathbb{K} = \mathbb{C}$, in the non-commutative case $\mathbb{K} = \mathbb{H}$ it is necessary to distinguish two types of vector spaces, namely, right vector spaces and left vector spaces.

We recall (see, e.g., [2]) that an additive abelian group X is a right $\mathbb{H}$-vector space if there is a map $X \times \mathbb{H} \to X$, under which the image of each pair $(x,q) \in X \times \mathbb{H}$ is denoted by xq, such that for all $q,q',q'' \in \mathbb{H}$ and $x,x',x'' \in X$,

- $(x' + x'')q = x'q + x''q$;
- $x(q' + q'') = xq' + xq''$;
- $x(q'q'') = (xq')q''$;
- $xe_0 = x$.

Similarly, an additive abelian group X is a left $\mathbb{H}$-vector space if there is a map $\mathbb{H} \times X \to X$, under which the image of each pair $(q,x) \in \mathbb{H} \times X$ is denoted by qx, such that for all $q,q',q'' \in \mathbb{H}$ and $x,x',x'' \in X$.

- $q(x' + x'') = qx' + qx''$;
- $(q' + q'')x = q'x + q''x$;
- $(q'q'')x = q'(q''x)$;
- (d') $e_0 x = x$.

According to these definitions, we define the left vector space $\mathbb{H}_l^n$ to be the space of row vectors $x = (x1,x2,...,xn)$, $xj \in \mathbb{H}$, with multiplication by scalars $c \in \mathbb{H}$ from the left-hand side ($x \to cx = (cx_1, cx_2, ..., cx_n)$). We equip $\mathbb{H}_l^n$ with the left inner product.

$$\langle x, y \rangle_l = \sum_{j=1}^{n} x_j \bar{y}_j. \tag{2.15}$$

The right vector space $\mathbb{H}_r^n$ is defined as the space of column vectors $x = (x_1, x_2, ..., x_n)^\mathsf{T}$, $xj \in \mathbb{H}$, with multiplication by scalars $c \in \mathbb{H}$ from the right-hand side ($x \to xc = (x1c, x2c, ..., xnc)^\mathsf{T}$) and with the right inner product.

$$\langle x, y \rangle_r = \sum_{j=1}^{n} \bar{x}_j y_j \tag{2.16}$$

Clearly, $\overline{\langle x, y \rangle l} = \langle y, x \rangle l, \overline{\langle x, y \rangle r} = \langle y, x \rangle r$. Furthermore, if $x^* = (\bar{x})^\mathsf{T}$, then

$$\langle x, y \rangle_l = \langle x^*, y^* \rangle_r, \qquad \langle x, y \rangle_r = \langle x^*, y^* \rangle_l.$$

If c is a real number, we can write $cx = xc$ for both $x \in \mathbb{H}_l^n$ and $x \in \mathbb{H}_r^n$. If $\mathbb{K} = \mathbb{C}$ (or $\mathbb{R}$) we regard $\mathbb{C}^n$ (or $\mathbb{R}^n$) as the space of column vectors and set

$$\langle x, y \rangle = \sum_{j=1}^{n} \bar{x}_j y_j \tag{2.17}$$

as in (2.16) (in the commutative case, definitions (2.15) and (2.16) coincide up to conjugation: $\langle x, y \rangle l = \overline{\langle x, y \rangle r}$).

Definition

We write $\mathbb{K}^n$ for the vector spaces $\mathbb{H}_l^n$, $\mathbb{H}_r^n$, $\mathbb{C}^n$, and $\mathbb{R}^n$, equipped with the inner product defined above.

There is a natural bijection h: $\mathbb{K}^n \to \mathbb{R}^N$, N = dn, where d = 1, 2, and 4 in the real, complex, and quaternionic cases, respectively. Specifically,

$$\mathbb{H}_l^n \ni x = (x_1, \ldots, x_n) \xrightarrow{h} v_x = \begin{bmatrix} v_{x_1} \\ \cdots \\ v_{x_n} \end{bmatrix} \in \mathbb{R}^{4n} \tag{2.18}$$

$$\mathbb{H}_r^n \ni x = \begin{bmatrix} x_1 \\ \cdots \\ x_n \end{bmatrix} \xrightarrow{h} v_x = \begin{bmatrix} v_{x_1} \\ \cdots \\ v_{x_n} \end{bmatrix} \in \mathbb{R}^{4n} \tag{2.19}$$

$$\mathbb{C}^n \ni x = (x_1, \ldots, x_n) \xrightarrow{h} v_x = \begin{bmatrix} v_{x_1} \\ \cdots \\ v_{x_n} \end{bmatrix} \in \mathbb{R}^{2n} \tag{2.20}$$

where $v_{x_i} = h(x_i)$. Abusing notation, we use the same letter h for both the scalar case, as in Section 2.2, and the vector case, as in (2.18)–(2.20).

Formulas (2.7) and (2.13) have obvious extensions in terms of block scalar matrices. Namely, for $x \in (\mathbb{H}^n)l$:

$$v_{qx} = \begin{bmatrix} v_{qx_1} \\ \cdots \\ v_{qx_n} \end{bmatrix} = \begin{bmatrix} L_q v_{x_1} \\ \cdots \\ L_q v_{x_n} \end{bmatrix} = \mathcal{L}_q v_x, \quad \mathcal{L}_q = \mathrm{diag}(L_q, \ldots, L_q) \tag{2.21}$$

for $x \in (\mathbb{H}^n)r$:

$$
v_{xq} = \begin{bmatrix} v_{x_1 q} \\ \cdots \\ v_{x_n q} \end{bmatrix} = \begin{bmatrix} R_q v_{x_1} \\ \cdots \\ R_q v_{x_n} \end{bmatrix} = \mathcal{R}_q v_x, \quad \mathcal{R}_q = \mathrm{diag}(R_q, \ldots, R_q)
$$

(2.22)

$$
\mathcal{J}\mathcal{L}_q\mathcal{J} = \mathcal{R}_{\bar{q}}, \qquad \mathcal{J}\mathcal{R}_q\mathcal{J} = \mathcal{L}_{\bar{q}}, \qquad \mathcal{J} = \mathrm{diag}(J, \ldots, J)
$$

(2.23)

Matrices $\mathcal{L}_q$, $\mathcal{R}_q$, and $J\mathcal{J}$ have n blocks; $\mathcal{L}_q$ and $\mathcal{R}_q$ belong to SO (4_n), and $\mathcal{J}^2$ is the identity matrix.

By (2.3), the inner product (2.15) can be written as

$$
\langle x, y \rangle_l = \sum_{i=0}^{3} \langle x, y \rangle_i e_i
$$

(2.24)

$$
\langle x, y \rangle_i = v_x \cdot \mathcal{A}_i v_y, \quad \mathcal{A}_i = \mathrm{diag}(A_i, \ldots, A_i) \quad (n \text{ blocks})
$$

(2.25)

A_i being defined by (2.4). Similarly, by (2.5),

$$
\langle x, y \rangle_r = \sum_{i=0}^{3} \langle x, y \rangle_i' e_i
$$

(2.26)

$$
\langle x, y \rangle_i' = v_x \cdot \mathcal{A}_i' v_y, \quad \mathcal{A}_i' = \mathrm{diag}(A_i', \ldots, A_i')
$$

(2.27)

By (2.11) and (2.23),

$$
\mathcal{J}\mathcal{A}_i\mathcal{J} = \mathcal{A}_i', \quad i = 0, 1, 2, 3
$$

(2.28)

In the case $\mathbb{K} = \mathbb{C}$, for $x \in \mathbb{C}^n$ and $c \in \mathbb{C}$, owing to (2.14), we have

$$v_{cx} = v_{xc} = \mathcal{M}_c v_x, \quad \mathcal{M}_c = \mathrm{diag}(\mathcal{M}_c, \ldots, \mathcal{M}_c) \in SO(2n)$$

(2.29)

Moreover,

$$\langle x, y \rangle = v_x \cdot v_y - i(v_x \cdot \mathcal{B} v_y)$$

(2.30)

$$\mathcal{B} = \mathrm{diag}\left(\begin{bmatrix} 0 & -1 \\ 1 & 0 \end{bmatrix}, \ldots, \begin{bmatrix} 0 & -1 \\ 1 & 0 \end{bmatrix} \right)$$

(2.31)

We introduce the following subgroups of block scalar matrices, consisting of n isoclinic blocks:

$$G_{\mathbb{H},l} = \left\{ g \in SO(4n) : g = \mathcal{L}_q = \mathrm{diag}(L_q, \ldots, L_q) \text{ for some } q \in \mathbb{H}, \ |q| = 1 \right\}$$

(2.32)

$$G_{\mathbb{H},r} = \left\{ g \in SO(4n) : g = \mathcal{R}_q = \mathrm{diag}(R_q, \ldots, R_q) \text{ for some } q \in \mathbb{H}, \ |q| = 1 \right\}$$

(2.33)

$$G_{\mathbb{C}} = \left\{ g \in SO(2n) : g = \mathcal{M}_c = \mathrm{diag}(M_c, \ldots, M_c) \text{ for some } c \in \mathbb{C}, \ |c| = 1 \right\}$$

(2.34)

If $=$, then the corresponding group $G_{\mathbb{R}}$ consists of two elements, namely, in and $-I_n$. The groups $G_{\mathbb{H},l}$ and $G_{\mathbb{H},r}$ are conjugate to each other by involution $\mathcal{J}$:

$$G_{\mathbb{H},l} = \mathcal{J} G_{\mathbb{H},r} \mathcal{J}$$

(2.35)

Definition

We will use the unified notation G for groups $G_{\mathbb{H},l}$, $G_{\mathbb{H},r}$, $G_{\mathbb{C}}$, and $G_{\mathbb{R}}$.

Equilibrated Convex Bodies

It is known that origin-symmetric convex bodies in $\mathbb{R}^n$ are in one-to-one correspondence with norms on $\mathbb{R}^n$. What is a natural analogue of this class of bodies in spaces over more general fields or algebras? Below we study this question in the general context of spaces over associative real normed algebras $\mathfrak{A}$ with identity. Our consideration generalizes the known reasoning for real and complex numbers [5, 19, 28,48].

We assume that $\mathfrak{A}$ contains real numbers and denote by $|\lambda|$ the norm of an element λ in $\mathfrak{A}$. Let V be a left (or right) module over $\mathfrak{A}$. By relating vectors in V new elements, called points, one obtains an affine space over $\mathfrak{A}$ [49]. We keep the same notation V for this affine space. As usual, a set A in V is called convex if $x \in A$ and $y \in A$ implies $\alpha x + \beta y \in A$ for all $\alpha \geq 0$, $\beta \geq 0$, $\alpha + \beta = 1$. A compact convex set in V with non-empty interior is called a convex body.

Definition

A set A in a left (right) space V over A is called equilibrated if for all $x \in A$, $\lambda x \in A$ ($x\lambda \in A$) whenever $\lambda \in$, $|\lambda| \leq 1$.

An equilibrated set in Rn is just an origin-symmetric star-shaped set. The next definition agrees with standard terminology for normed algebras; cf. [31, p. 655].

Definition

Let V be a left space over $\mathfrak{A}$. A function $p \colon V \to \mathbb{R}$ is called a norm if the following conditions are satisfied:

- $p(x) \geq 0$ for all $x \in V$; $p(x) = 0$ if and only if $x = 0$;

- $p(\lambda x) = |\lambda| p(x)$ for all $x \in V$ and all $\lambda \in$;

- $p(x + y) \leq p(x) + p(y)$ for all $x, y \in V$.

If V is a right space over then (b) is replaced by

$(b')p(x\lambda)=|\lambda|p(x)$ for all $x \in V$ and all $\lambda \in \mathfrak{A}$.

Let $V = {}^n$ be the n-dimensional left (right) affine space over $\mathfrak{A}$. Every point $x \in V$ is represented as $x = x_1 f_1 + \cdots + x_n f_n$ ($x = f_1 x_1 + \cdots + f_n x_n$), where $xi \in \mathfrak{A}$ and $f_1 = (1, 0,..., 0),...,f_n = (0, 0,..., 1)$ is a standard basis in V.

We set $\|x\|_2 = \left(\sum_{i=1}^{n}|x_i|^2\right)^{1/2}$.

Lemma

Let $V = \mathfrak{A}^n$ be a left (right) space over $\mathfrak{A}$.

- If $p: V \to \mathbb{R}$ is a norm, then

$$A_p = \{x \in V: \ p(x) \leqslant 1\}$$

(2.37)

- Conversely, if A is an equilibrated convex body in V, then

is a norm in V such that $A = \{x \in V : \|x\| \leq 1\}$.

The proof of this lemma is standard and can be found in [58].

In the following $\equiv\ \in \{\ ,\ \};\ ^n$ is any of the spaces $^n, {}^n\ \mathbb{H}_l^n$ or $\mathbb{H}_r^n$ $G \in \{$ $G_\mathbb{R}, G_\mathbb{C}, G_\mathbb{H}, l, G_\mathbb{H}, r\}$; see Definitions 2.2 and 2.3; $N = n, 2n$, or $4n$, respectively. Our next aim is to establish a connection between equilibrated convex bodies in ^{n} and G-invariant originsymmetric star bodies in $^n = h(^n)$. We recall the notation

$$\mathcal{J} = \text{diag}\left(\begin{bmatrix} -1 & 0 \\ 0 & I_3 \end{bmatrix},\cdots,\begin{bmatrix} -1 & 0 \\ 0 & I_3 \end{bmatrix}\right) \quad (n \text{ blocks}).$$

(2.38)

Clearly, J acts on $\xi = (\xi_1,\xi_2,...,\xi_{4n}) \in {}^{4n}$ by converting ξ_1 into $-\xi_1$, ξ_5 into $-\xi_5$, and so on.

Theorem

Let A be a set in ^{n} and let $B = h(A)$ be its image in $\mathbb{R}^N$. Then

- A is convex if and only if B is convex.
- A is equilibrated in $\mathbb{H}_l^n$ if and only if B is $G_{\mathbb{H}}$,l-invariant and star-shaped.
- A is equilibrated in $\mathbb{H}_r^n$ if and only if B is $G_{\mathbb{H}}$,r -invariant and star-shaped.
- A is equilibrated in $\mathbb{C}^n$ if and only if B is $G_{\mathbb{C}}$ -invariant and star-shaped.
- A is equilibrated in $\mathbb{R}^n$ if and only if it is origin-symmetric and star-shaped.
- A set S in $\mathbb{R}^{4n}$ is star-shaped and $G_{\mathbb{H}}$, l -invariant (or $G_{\mathbb{H}}$, r -invariant) if and only if the reflected set J S is star-shaped and $G_{\mathbb{H}}$, r -invariant ($G_{\mathbb{H}}$, l -invariant, respectively).

The proof is straightforward; see [58] for details.

Central Hyperplanes in n and the G-invariant Busemann–Petty Problem in $\mathbb{R}^N$

Let $S_{\mathbb{K}^n} = \left\{ y \in \mathbb{K}^n l : \|y\|_2 = 1 \right\}$ be the unit sphere in $\mathbb{K}^n$. Every hyperplane in $\mathbb{K}^n$ passing through the origin has the form

$$y^{\perp} = \left\{ x \in \mathbb{K}^n : \langle x, y \rangle = 0 \right\}, \quad y \in S_{\mathbb{K}^n} \tag{2.39}$$

where x,y is the relevant inner product; see (2.15), (2.16), (2.17).

If $\mathbb{K} = \mathbb{R}$, this is a usual (n−1)-dimensional subspace of $\mathbb{R}^n$. If $\mathbb{K} = \mathbb{C}$, then, owing to (2.30), the equality $\langle x, y \rangle_1 = 0$ is equivalent to a system of two equations

$$\xi \cdot \theta = 0, \quad \xi \cdot B\theta = 0$$

where $\xi = h(x) \in \mathbb{R}^{2n}$, $\theta = h(y) \in S^{2n-1}$, B being defined by (2.31). This system can be replaced by one matrix equation

$$F_2(\theta)^T \xi = 0, \qquad F_2(\theta) = [\theta, B\theta] \in V_{2n,2} \tag{2.40}$$

where $V_{2n,2}$ is the Stiefel manifold of orthonormal 2-frames in $\mathbb{R}^{2n}$. Eq. (2.40) defines a $(2n - 2)$-dimensional subspace of $\mathbb{R}^{2n}$. The collection of all such subspaces will be denoted by $\mathrm{Gr}^{\mathbb{C}}_{2n-2}\left(\mathbb{R}^{2n}\right)$.

In the non-commutative case $\mathbb{K} = \mathbb{H}$ we have two options. If $\mathbb{K}^n = \mathbb{H}^n_l$, then, owing to (2.24), the equality $\langle x, y \rangle_l = 0$ is equivalent to a system of four equations

$$\xi \cdot A_i \theta = 0 \quad (i = 0, 1, 2, 3)$$

or

$$F_{4,l}(\theta)^T \xi = 0, \qquad F_{4,l}(\theta) = [A_0\theta, A_1\theta, A_2\theta, A_3\theta] \in V_{4n,4}, \tag{2.41}$$

where $\xi = h(x) \in \mathbb{R}^{4n}$, and $\theta = h(y) \in S^{4n-1}$ (for simplicity, we use the same letters). If $\mathbb{K}^n = \mathbb{H}^n_r$, then, by (2.26), $\langle x, y \rangle_r = 0$ is equivalent to

$$F_{4,r}(\theta)^T \xi = 0, \qquad F_{4,r}(\theta) = \left[A'_0\theta, A'_1\theta, A'_2\theta, A'_3\theta\right] \in V_{4n,4} \tag{2.42}$$

Since $A'_i = \mathcal{J} A_i \mathcal{J}$ (see (2.28)), then

$$F_{4,r}(\theta) = \mathcal{J} F_{4,l}(\mathcal{J}\theta) \quad \text{for every } \theta \in S^{4n-1} \tag{2.43}$$

Thus, (2.41) and (2.42) define two different $(4n-4)$-dimensional sub-spaces of $\mathbb{R}^{4n}$ generated by the same point $\theta \in S^{4n-1}$. We denote by $Gr^{H,l}_{4n-4}(\mathbb{R}^{4n})$ and GrH,r $4n-4(\mathbb{R}^{4n})$ respective collections of all such subspaces, which are isomorphic to S4n−1. By (2.43),

$$Gr^{H,r}_{4n-4}(\mathbb{R}^{4n}) = \mathcal{J}\,Gr^{H,l}_{4n-4}(\mathbb{R}^{4n})$$

Given $\theta \in S^{dn-1}$ $(d = 1, 2, 4)$, we will be using the unified notation Hθ for the $(dn - d)$- dimensional subspace orthogonal to $F_1(\theta) = \theta$, $F_2(\theta)$, $F_{4,l}(\theta)$, and $F_{4,r}(\theta)$, respectively.

Proposition

The "right" manifold $G^{H,r}_{4n-4}(\mathbb{R}^{4n})$ is invariant under the "left" rotations $\mathcal{L}_q$, that is

$$\mathcal{L}_q\,Gr^{H,r}_{4n-4}(\mathbb{R}^{4n}) = Gr^{H,r}_{4n-4}(\mathbb{R}^{4n})$$

The "left" manifold $G^{H,l}_{4n-4}(\mathbb{R}^{4n})$ is invariant under the "right" rotations $\mathcal{R}_q$, that is

$$\mathcal{R}_q\,Gr^{H,l}_{4n-4}(\mathbb{R}^{4n}) = Gr^{H,l}_{4n-4}(\mathbb{R}^{4n})$$

Proof: Let $H \in G^{H,r}_{4n-4}(\mathbb{R}^{4n})$, that is, H is orthogonal to $F_{4,r}(\theta) = [\mathcal{A}'_0\theta, \mathcal{A}'_1\theta, \mathcal{A}'_2\theta, \mathcal{A}'_3\theta,]$ for some $\theta \in S_{4n-1}$. Since $\mathcal{L}_p$ and $\mathcal{R}_q$ commute for any $p,q \in \mathbb{H}$ and $\mathcal{A}'_i = \mathcal{R}_{\bar{e}_i}$ (see (2.12) and (2.11)), then $\mathcal{L}_q\mathcal{A}'_i = \mathcal{A}'_i\mathcal{L}_q$ and $\mathcal{L}_q F_{4,r}(\theta) F_{4,r}(\mathcal{L}_q\theta)$ This implies

$$\mathcal{L}_q \, \mathrm{Gr}^{\mathbb{H},r}_{4n-4}\!\left(\mathbb{R}^{4n}\right) \subset \mathrm{Gr}^{\mathbb{H},r}_{4n-4}\!\left(\mathbb{R}^{4n}\right)$$

for the corresponding bundles of subspaces. By the same reason, we have $F_{4,r}(\theta) = \mathcal{L}_q F_{4,r}\!\left(\mathcal{L}_q^{-1}\theta\right)$ which gives the opposite embedding. The proof of equality $\mathcal{R}_q \mathrm{G}^{\mathbb{H},l}_{4n-4}\!\left(\mathbb{R}^{4n}\right) = \mathrm{G}^{\mathbb{H},l}_{4n-4}\!\left(\mathbb{R}^{4n}\right)$ is similar.

The above consideration enables us to give precise setting of the hyperplane slice (HS) comparison problem of the Busemann–Petty type in $\mathbb{K}^n$ and reformulate the latter as the equivalent lower dimensional problem for G-invariant convex bodies in $\mathbb{R}^N$. We recall that

$$N = dn, \quad n > 1, \ d = 1, 2, 4; \qquad G \in \{G_{\mathbb{R}}, G_{\mathbb{C}}, G_{\mathbb{H},l}, G_{\mathbb{H},r}\}$$

see (2.32)–(2.34). We will be using the unified notation $\widetilde{G}_{rN-d}\,(\mathbb{R}^N)$ for the respective manifolds

$$\mathrm{Gr}_{n-1}\!\left(\mathbb{R}^n\right), \quad \mathrm{Gr}^{\mathbb{C}}_{2n-2}\!\left(\mathbb{R}^{2n}\right), \quad \mathrm{Gr}^{\mathbb{H},l}_{4n-4}\!\left(\mathbb{R}^{4n}\right), \quad \mathrm{Gr}^{\mathbb{H},r}_{4n-4}\!\left(\mathbb{R}^{4n}\right)$$

of $(N - d)$-dimensional subspaces H_θ introduced above.

The Hyperplane Slice Comparison Problem (Problem A)

Let A and B be equilibrated convex bodies in $\mathbb{K}^n$, $n > 1$, satisfying

$$\mathrm{vol}_{n-1}(A \cap \xi) \leqslant \mathrm{vol}_{n-1}(B \cap \xi) \tag{2.44}$$

for all central K-hyperplanes ξ. Does it follow that $\mathrm{vol}_n(A) \leq \mathrm{vol}_n(B)$?

Here volumes of geometric objects in $\mathbb{K}^n$ are defined as usual volumes of their h-images

in $\mathbb{R}^N$, for example,

$$\mathrm{vol}_n(A) = \mathrm{vol}_N\big(h(A)\big), \qquad \mathrm{vol}_{n-1}(A \cap \xi) = \mathrm{vol}_{N-d}\big(h(A \cap \xi)\big)$$

An equivalent lower dimensional slice (LDS) comparison problem is formulated as follows.

The LDS Comparison Problem (Problem B)

Let K and L be G-invariant convex bodies in $\mathbb{R}^N$, with section functions

$$S_K(\theta) = \mathrm{vol}_{N-d}(K \cap H_\theta), \qquad S_L(\theta) = \mathrm{vol}_{N-d}(L \cap H_\theta)$$

where $H_\theta \in \widetilde{G}_{N-d}\big(\mathbb{R}^N\big)$. Suppose that $S_K(\vartheta) \le S_L(\vartheta)$ for all $\vartheta \in S^{N-1}$. Does it follow that $\mathrm{vol}_N(K) \le \mathrm{vol}_N(L)$?

We notice a fundamental difference between the usual lower dimensional Busemann–Petty problem, where sections by all $(N - d)$-dimensional subspaces are compared, and Problem B, where, in the cases d = 2 and 4, the essentially smaller (actually, $(N-1)$-dimensional) collection of subspaces comes into play.

Since the question in Problem B may have a negative answer, we also consider the following more general problem, which is of independent interest.

The Generalized LDS Comparison Problem (Problem C)

Let K, L, $S_K(\vartheta)$, and $S_L(\vartheta)$ have the same meaning as in Problem B. For which operator D does the assumption $DS_K(\vartheta) \le DS_L(\vartheta)$, $\forall \vartheta \in S^{N-1}$ imply $\mathrm{vol}_N(K) \le \mathrm{vol}_N(L)$?

Vector Fields on Spheres

Theorem 2.1 suggests intriguing links between possible generalizations of Problems B and C and the celebrated vector field problem, which asks for the maximal number $\rho(d)$ of orthonormal tangent vector fields on the unit sphere S^{d-1} in $\mathbb{R}^d$.

We recall some facts; see [27,30,1]. A continuous tangent vector field on S^{d-1} is defined to be a continuous function $V : S^{d-1} \to \mathbb{R}^d$ such that $V(\sigma) \in \sigma^{\perp}$ for every $\sigma \in S^{d-1}$. If $V(\sigma) = A\sigma$, where A is a $d \times d$ matrix, the vector field V is called linear. Vector fields $V_1, \ldots, V_k$ on $Sd-1$ are called orthonormal if for every $\sigma \in S^{d-1}$, the corresponding vectors $V_1(\sigma), \ldots, V_k(\sigma)$ form an orthonormal frame in $\mathbb{R}^d$. The following result is known as the Hurwitz–Radon–Eckmann theorem [29,47,12]; see also [42].

Theorem

Let d be a positive integer and write $d = 2^{4s+r} t$, where t is an odd integer, s and r are integers with $s \geq 0$ and $0 \leq r < 4$. Then the maximal number of orthonormal linear tangent vector fields on S^{d-1} is equal to $\rho(d) = 2^r + 8s - 1$.

The number $\rho(d)$ is called the Radon–Hurwitz number. It is zero when d is odd. In a groundbreaking paper, Adams [1] extended this result to continuous vector fields. He proved that there are at most $\rho(d)$ linearly independent continuous tangent vector fields on S^{d-1}.

In the case $\rho(d) = d - 1$, when there exists a complete orthonormal system of linear tangent vector fields $\{V_1, \ldots, V_{d-1}\}$ on S^{d-1}, the sphere S^{d-1} is called parallelizable. The only parallelizable spheres are S^1, S^3, and S^7; see Kervaire [32], Bott and Milnor [7].

Complete systems of orthonormal linear tangent vector fields on S^3, namely, $\{A_1 \sigma, A_2 \sigma, A_3 \sigma\}$ and $\{A'_1\sigma, A'_2\sigma, A'_3\sigma\}$ where considered in Theorem 2.1. These produce a series of new examples, for instance,

$$\{[\gamma^{-1} A_1 \gamma]\sigma, [\gamma^{-1} A_2 \gamma]\sigma, [\gamma^{-1} A_3 \gamma]\sigma\}, \quad \forall \gamma \in O(4) \tag{2.45}$$

Example: A complete system of orthonormal tangent linear vector fields on S^7 can be constructed, e.g., as follows.

If $\sigma = (\sigma_1, \sigma_2, \sigma_3, \sigma_4, \sigma_5, \sigma_6, \sigma_7, \sigma_8)^{\top} \in S^7$, then

$$A_1\sigma = (\sigma_2, -\sigma_1, \sigma_4, -\sigma_3, \sigma_6, -\sigma_5, -\sigma_8, \sigma_7)^T,$$

$$A_2\sigma = (\sigma_3, -\sigma_4, -\sigma_1, \sigma_2, \sigma_7, \sigma_8, -\sigma_5, -\sigma_6)^T,$$

$$A_3\sigma = (\sigma_4, \sigma_3, -\sigma_2, -\sigma_1, \sigma_8, -\sigma_7, \sigma_6, -\sigma_5)^T,$$

$$A_4\sigma = (\sigma_5, -\sigma_6, -\sigma_7, -\sigma_8, -\sigma_1, \sigma_2, \sigma_3, \sigma_4)^T,$$

$$A_5\sigma = (\sigma_6, \sigma_5, -\sigma_8, \sigma_7, -\sigma_2, -\sigma_1, -\sigma_4, \sigma_3)^T,$$

$$A_6\sigma = (\sigma_7, \sigma_8, \sigma_5, -\sigma_6, -\sigma_3, \sigma_4, -\sigma_1, -\sigma_2)^T,$$

$$A_7\sigma = (\sigma_8, -\sigma_7, \sigma_6, \sigma_5, -\sigma_4, -\sigma_3, \sigma_2, -\sigma_1)^T.$$

The corresponding matrices A_i, which are determined by permutation of indices of coordinates $\sigma_1, \ldots, \sigma_8$ and arrangements of $\pm$ signs, belong to SO(8). More systems can be constructed, e.g., as in (2.45).

The following statement can be found in [27] in a slightly more general form. For the sake of completeness, we present it with proof.

Lemma

- If $\sigma \to A\sigma$ is a linear tangent vector field on Sd−1, then the d ×d matrix A is skew symmetric, that is, $A + A^T = 0$.
- If $A\sigma = \{A_i\sigma\}_{i=1}^{d-1}$ is an orthonormal system of linear tangent vector fields on S^{d-1}, then

$$A_i^T A_j + A_j^T A_i = 0 \quad \text{for all } 1 \leqslant i < j \leqslant d - 1$$

$$A_i^T A_i = I \quad \text{for all } 1 \leqslant i \leqslant d - 1.$$

Proof

- Let $\sigma \cdot A\sigma = 0$ for all $\sigma \in S^{d-1}$. Equivalently, $x \cdot Ax = 0$ for all $x \in \mathbb{R}^d$. Then, for all $x, y \in \mathbb{R}^d$,

$$x \cdot (A + A^T) y = x \cdot Ay + Ax \cdot y$$

$$= x \cdot Ax + x \cdot Ay + Ax \cdot y + Ay \cdot y = (x + y) \cdot A(x + y) = 0.$$

Hence, $A + A^T = 0$.

- (ii) As above, for all $x, y \in \mathbb{R}^d$ we have

$$x \cdot \left(A_i^T A_j + A_j^T A_i\right) y = A_i(x+y) \cdot A_j(x+y) = 0,$$

$$x \cdot \left(A_i^T A_i - I\right) y = \frac{1}{2}\left[A_i(x+y) \cdot A_i(x+y) - (x+y) \cdot (x+y)\right] = 0$$

This gives the result.

Lemma

Let $A\sigma = \{A_i\sigma\}_{i=1}^{d-1}$ be an orthonormal system of linear tangent vector fields on S^{d-1}; $A_0 = I$. Then

$$g_\lambda(\mathbf{A}) \equiv \sum_{i=0}^{d-1} \lambda_i A_i \in O(d)$$

(2.46)

for every $\lambda = (\lambda_1, \ldots, \lambda_n) \in S^{d-1}$.

Proof

By Lemma 2.11,

$$g_\lambda(\mathbf{A})^T g_\lambda(\mathbf{A}) = \left(\sum_{i=0}^{d-1} \lambda_i A_i^T\right)\left(\sum_{j=0}^{d-1} \lambda_j A_j\right) = \sum_{i,j=0}^{d-1} \lambda_i \lambda_j A_i^T A_j = I$$

Hence, $g_\lambda(A) \in O(d)$.

Some notations are in order.

Definition

Let $N = dn$, $d \in \{2, 4, 8\}$, $n > 1$. Given a complete orthonormal system $\sigma = \{A_i \sigma\}_{i=1}^{d-1}$ of linear tangent vector fields on S^{d-1}, we introduce the block scalar matrices

$$\mathcal{G}_\lambda(\mathbf{A}) = \mathrm{diag}\left(g_\lambda(\mathbf{A}), \ldots, g_\lambda(\mathbf{A})\right)$$

$$= \mathrm{diag}\left(\sum_{i=0}^{d-1} \lambda_i A_i, \ldots, \sum_{i=0}^{d-1} \lambda_i A_i\right) \quad (n \text{ equal blocks})$$

$$(2.47)$$

where $\lambda \in S^{d-1}$. The corresponding class of block diagonal orthogonal transformations of $\mathbb{R}^N$ (with n equal d ×d diagonal blocks), generated by A, is defined by

$$G \equiv G(n, d; \mathbf{A}) = \left\{g \in O(N): g = \mathcal{G}_\lambda(\mathbf{A}) \text{ for some } \lambda \in S^{d-1}\right\}$$

$$(2.48)$$

We also introduce N ×N block scalar matrices, containing n blocks:

$$\mathcal{A}_i = \mathrm{diag}(A_i, \ldots, A_i) \quad (i = 1, 2, \ldots, d-1)$$

$$(2.49)$$

and set A0 = IN. Given $\vartheta \in S^{N-1}$, we denote by H_ϑ the (N − d)-dimensional subspace orthogonal to the d-frame

$$F_d(\theta) = [\theta, \mathcal{A}_1\theta, \ldots, \mathcal{A}_{d-1}\theta] \in V_{N,d}$$

$$(2.50)$$

and set

$$\widetilde{\mathrm{Gr}}_{N-d}(\mathbb{R}^N) = \left\{H_\theta: \theta \in S^{N-1}\right\}$$

$$(2.51)$$

All objects in Definition 2.13 are familiar to us when d = 2, 4 (see Section 2.5). Thus, Problems B and C extend to the case d = 8.

We recall that the set G of transformations and the set $\widetilde{Gr}_{N-d}\left(\mathbb{R}^N\right)$ of planes are determined by the orthonormal system $A = \{A_i\sigma\}_{i=1}^{d-1}$ of vector fields, which is assumed to be fixed.

The following lemma plays a crucial role in our consideration.

Lemma

If $H \in \widetilde{Gr}_{N-d}\left(\mathbb{R}^N\right)$, then every continuous G-invariant function f on S^{N-1} is constant on the (d −1)-dimensional section $S^{N-1} \cap H^{\perp}$.

Proof. Let $H \equiv H_\vartheta$ be orthogonal to some d-frame (2.50). Any point $\eta \in S^{N-1} \cap H^{\perp}$ is represented as

$$\eta = \sum_{i=0}^{d-1} \lambda_i A_i \theta, \qquad \sum_{i=0}^{d-1} \lambda_i^2 = 1$$

or $\eta = \mathcal{G}_\lambda(A)\vartheta$; see (2.47). In particular, if d = 4 and Ai have the form (2.4), then $\mathcal{G}_\lambda$ (A) is a block diagonal matrix with n equal blocks of the form

$$\sum_{i=0}^{3} \lambda_i A_i = \begin{bmatrix} \lambda_0 & -\lambda_1 & -\lambda_2 & -\lambda_3 \\ \lambda_1 & \lambda_0 & -\lambda_3 & \lambda_2 \\ \lambda_2 & \lambda_3 & \lambda_0 & -\lambda_1 \\ \lambda_3 & -\lambda_2 & \lambda_1 & \lambda_0 \end{bmatrix} = L_\lambda.$$

$$\lambda = \lambda_0 e_0 + \lambda_1 e_1 + \lambda_2 e_2 + \lambda_3 e_3 \in \mathbb{H} \quad (\text{cf. (2.8)})$$

Since $\mathcal{G}_\lambda$ (A) $\in$ G, then f (η) = f $(\mathcal{G}_\lambda$ (A) $\theta)$ = f (ϑ). This gives the result.

COSINE TRANSFORMS AND INTERSECTION BODIES

It is known [53,54,56,59] that diverse Busemann–Petty type problems can be studied using analytic families of cosine transforms on the unit sphere. This approach is parallel, in a sense, to the Fourier transform method developed by Koldobsky and his collaborators [33,35]. We shall see how these transforms can be applied to Problems A, B, and C stated above.

Spherical Radon Transforms and Cosine Transforms

We recall some basic facts; see [52,56]. Fix an integer $i \in \{2, 3, \ldots , N - 1\}$ and let $Gr_i(\mathbb{R}^N)$ be the Grassmann manifold of all i-dimensional linear subspaces ξ of $\mathbb{R}^N$. The spherical Radon transform, that integrates a function $f \in L^1(S^{N-1})$ over $(i-1)$-dimensional sections $S^{N-1} \cap \xi$, is defined by

$$(R_i f)(\xi) = \int_{\theta \in S^{N-1} \cap \xi} f(\theta) \, d_\xi \theta$$

(3.1)

where $d_\xi \vartheta$ denotes the probability measures on $S^{N-1} \cap \xi$. The case $i = N - 1$ in (3.1) is known as the Minkowski–Funk transform

$$(Mf)(u) = \int_{\{\theta:\, \theta \cdot u = 0\}} f(\theta) \, d_u \theta = (R_{N-1} f)(u^\perp), \quad u \in S^{N-1}$$

(3.2)

Transformation (3.1) can be regarded as a member (up to a multiplicative constant) of the analytic family of the generalized cosine transforms

$$(R_i^\alpha f)(\xi) = \gamma_{N,i}(\alpha) \int_{S^{N-1}} |\text{Pr}_\xi \theta|^{\alpha+i-N} f(\theta) \, d\theta.$$

$$\gamma_{N,i}(\alpha) = \frac{\sigma_{N-1} \Gamma((N-\alpha-i)/2)}{2\pi^{(N-1)/2} \Gamma(\alpha/2)}, \quad \text{Re}\,\alpha > 0, \ \alpha + i - N \neq 0, 2, 4, \ldots.$$

(3.3)

Here $\mathrm{Pr}_{\xi^{\perp}}\vartheta$ stands for the orthogonal projection of ϑ onto $\xi^{\perp}$. If f is smooth and $\mathrm{Re}\,\alpha \leq 0$, then $R_i^{\alpha}f$ is understood as analytic continuation of the integral (3.3), so that

$$\lim_{\alpha \to 0} R_i^{\alpha} f = R_i^0 f = c_i\, R_i f, \quad c_i = \frac{\sigma_{i-1}}{2\pi^{(i-1)/2}}$$

(3.4)

In the case $i = N - 1$ we also set

$$(M^{\alpha} f)(u) = (R_{N-1}^{\alpha} f)(u^{\perp}) = \gamma_N(\alpha) \int_{S^{N-1}} f(\theta)|\theta \cdot u|^{\alpha-1}\, d\theta$$

(3.5)

$$\gamma_N(\alpha) = \frac{\sigma_{N-1}\Gamma((1-\alpha)/2)}{2\pi^{(N-1)/2}\Gamma(\alpha/2)}, \quad \mathrm{Re}\,\alpha > 0, \ \alpha \neq 1, 3, 5, \ldots$$

(3.6)

Lemma

(See [56, Lemma 3.2].) Let $\alpha, \beta \in$; $\alpha, \beta \neq 1, 3, 5, \ldots$ If $\alpha + \beta = 2 - N$ and $f \in_e (S^{N-1})$ then

$$M^{\alpha} M^{\beta} f = f$$

(3.7)

If α, $2 - N - \alpha \neq 1, 3, 5, \ldots$, then $M\alpha$ is an automorphism of $_e (S^{N-1})$.

Corollary

The Minkowski–Funk transform on the space $_e (S^{N-1})$ can be inverted by then formula

$$(M)^{-1} = c_{N-1} M^{2-N}, \quad c_{N-1} = \frac{\sigma_{N-2}}{2\pi^{(N-2)/2}}$$

(3.8)

Both statements amount to results of Semyanistyi [64], who used the Fourier transform techniques. They can also be obtained as immediate consequence of the spherical harmonic decomposition of $M^\alpha f$.

Lemma

(See [56, Lemma 3.5].) Let Re $\alpha > 0$; $\neq 1, 3, 5, \ldots$. If $f \in L^1(S^{N-1})$, then

$$(R_i M^\alpha f)(\xi) = c\, (R_{N-i}^{\alpha+i-1} f)(\xi^\perp), \quad \xi \in \mathrm{Gr}_i(\mathbb{R}^N), \quad c = \frac{2\pi^{(i-1)/2}}{\sigma_{i-1}} \tag{3.9}$$

$$(R_{N-i} M^\alpha f)(\xi^\perp) = \frac{2\pi^{(N-i-1)/2}}{\sigma_{N-i-1}} (R_i^{\alpha+N-i-1} f)(\xi). \tag{3.10}$$

If $f \in {}_e(S^{N-1})$, then (3.9) and (3.10) extend to Re $\alpha \le 0$ by analytic continuation.

Proof

We sketch the proof for the sake of completeness. For Re $\alpha > 0$,

$$(R_i M^\alpha f)(\xi) = \gamma_N(\alpha) \int_{S^{N-1} \cap \xi} d_\xi u \int_{S^{N-1}} f(\theta) |\theta \cdot u|^{\alpha-1}\, d\theta$$

Since $|\vartheta \cdot u| = |\mathrm{Pr}_\xi \vartheta||v\vartheta \cdot u|$ for some $v\vartheta \in S^{N-1} \cap \xi$, changing the order of integration, we obtain

$$(R_i M^\alpha f)(\xi) = \gamma_N(\alpha) \int_{S^{N-1}} f(\theta) |\mathrm{Pr}_\xi \theta|^{\alpha-1}\, d\theta \int_{S^{N-1} \cap \xi} |v_\theta \cdot u|^{\alpha-1}\, d_\xi u$$

The inner integral is independent of v_ϑ and can be easily evaluated. This gives (3.9). Equality (3.10) is a reformulation of (3.9).

An origin-symmetric star body K in $\mathbb{R}^N$ is completely determined by its radial function $\rho_K(\vartheta) = \sup\{\lambda \ge 0 : \lambda\vartheta \in K\}$; see notation in Section 2.1. Passing to polar coordinates, we get

$$\text{vol}_i(K \cap \xi) = \frac{\sigma_{i-1}}{i}(R_i \rho_K^i)(\xi), \quad \xi \in \text{Gr}_i(\mathbb{R}^N)$$

(3.11)

The next statement follows from Lemma 2.14 and plays the key role in the whole paper.

Lemma

Let $\rho_K \in D_e^G(S^{N-1})$, $N = dn$, $d \in \{1, 2, 4, 8\}$, $n > 1$. Then for every subspace $H\vartheta \in \widetilde{\text{Gr}}^{N-d}(\mathbb{R}^N)$ with $\vartheta \in S^{N-1}$,

$$\text{vol}_{N-d}(K \cap H_\theta) = \frac{\pi^{N/2-d}\sigma_{d-1}}{N-d}(M^{1-d}\rho_K^{N-d})(\theta)$$

(3.12)

Proof

Applying successively (3.11) (with $k = N - d$), (3.4), and (3.10) (with $\alpha = i + 1 - N$, $i = N - d$), we obtain

$$\text{vol}_{N-d}(K \cap H_\theta) = \frac{\sigma_{N-d-1}}{N-d}(R_{N-d}\rho_K^{N-d})(H_\theta)$$

$$= \frac{2\pi^{(N-d-1)/2}}{N-d}(R_{N-d}^0 \rho_K^{N-d})(H_\theta)$$

$$= \frac{\pi^{N/2-d}\sigma_{d-1}}{N-d}(R_d M^{1-d}\rho_K^{N-d})(H_\theta^\perp)$$

Since ρK is G-invariant and M^{1-d} commutes with orthogonal transformations, then, by

Lemma 2.14, $M^{1-d}\rho_K^{N-d} \equiv$ const on $S^{N-1} \cap H^\perp \vartheta$ and (3.12) follows.

Remark

In the classical case $\mathbb{K} = \mathbb{R}$, when $N = n$ and $d = 1$, (3.12) becomes a particular case of (3.11):

$$\mathrm{vol}_{n-1}(K \cap \theta^{\perp}) = \frac{\sigma_{n-2}}{n-1}(M\rho_K^{n-1})(\theta)$$

where M is the Minkowski–Funk transform (3.2).

Homogeneous Distributions and Riesz Fractional Derivatives

Given a G-invariant infinitely smooth body K in $\mathbb{R}^N$ and a plane $H\vartheta \in \widetilde{\mathrm{Gr}}^{N-d}(\mathbb{R}^N)$ generated by $\vartheta \in S^{N-1}$, we denote

$$S_K(\theta) = \mathrm{vol}_{N-d}(K \cap H_\theta) \tag{3.13}$$

Question

For which operator A^α,

$$A^\alpha M^{1-d} \rho_K^{N-d} = \left(M^{1-\alpha} \rho_K^{N-d}\right)(\theta)? \tag{3.14}$$

By (3.12), an answer to this question would give us the corresponding equality for the section function

$$A^\alpha S_K(\theta) = c\left(M^{1-\alpha} \rho_K^{N-d}\right)(\theta), \quad c = \frac{\pi^{N/2-d} \sigma_{d-1}}{N-d} \tag{3.15}$$

$$A^\alpha = M^{1-\alpha} M^{1+d-N} \tag{3.16}$$

To make this explicit formula more transparent and convenient to handle, we extend our functions by homogeneity to the entire space RN and invoke powers of the Laplacian. This idea was formally used in

[36,34], however, it requires a proper justification. Below we explain the essence of the matter.

Let $S(^N)$ be the Schwartz space of rapidly decreasing C^∞ functions, and $S'(^N)$ its dual. The Fourier transform of a distribution F in $S'(^N)$ is defined by[2]

$$\langle \hat{F}, \hat{\phi} \rangle = (2\pi)^N \langle F, \phi \rangle. \qquad \hat{\phi}(y) = \int_{\mathbb{R}^N} \phi(x) e^{ix \cdot y} \, dx, \quad \phi \in S(\mathbb{R}^N)$$

For $f \in L^1(S^{N-1})$, let

$$(E_\lambda f)(x) = |x|^\lambda f(x/|x|), \quad x \in \mathbb{R}^N \setminus \{0\}$$

This operator generates a meromorphic S'distribution, which is defined by analytic continuation (a.c.) as follows:

$$\langle E_\lambda f, \phi \rangle = a.c. \int_0^\infty r^{\lambda+N-1} u(r) \, dr, \quad u(r) = \int_{S^{N-1}} f(\theta) \overline{\phi(r\theta)} \, d\theta$$

The distribution $E_\lambda f$ is regular if $\mathrm{Re}\lambda > -N$ and admits simple poles at $\lambda = -N, -N-1, \ldots$; see [17]. If f is orthogonal to all spherical harmonics of degree j , then the derivative $u^{(j)}(r)$ equals zero at r = 0 and the pole at $\lambda = -N - j$ is removable. In particular, if f is even, that is, for every $\varphi \in (S^{N-1})$ we have $(f, \varphi) = (f, \varphi\_)$, where $\varphi-(\vartheta) \equiv \varphi(-\vartheta)$, then the only possible poles of $E_\lambda f$ are $-N, -N-2, -N-4, \ldots$.

The operator family $\{M^\alpha\}$ (see (3.5)) naturally arises thanks to the formula

$$[E_{1-N-\alpha} f]^\wedge = 2^{1-\alpha} \pi^{N/2} E_{\alpha-1} M^\alpha f, \quad f \in D_e(S^{N-1})$$

$$(3.17)$$

which amounts to Semyanistyi [64]. It holds pointwise for $0 < \text{Re } \alpha < 1$ (see, e.g., Lemma 3.3 in [51]) and extends in the S'sense to all $\alpha \in$ satisfying

$$\alpha \notin \{1, 3, 5, \ldots\} \cup \{1 - N, -N - 1, -N - 3, \ldots\}$$

(3.18)

The Riesz fractional derivative $D^{\alpha}\psi$ of order $\alpha \in$ of a Schwartz function ψ is defined as an S' (RN)-distribution by the rule

$$(2\pi)^N \langle D^{\alpha}\psi, \phi \rangle = \langle |y|^{\alpha}\hat{\psi}, \hat{\phi} \rangle, \quad \phi \in S(\mathbb{R}^N)$$

(3.19)

where the right-hand side is a meromorphic function of α with simple poles $\alpha = -N, -N - 2, \ldots$. One can formally regard D^{α} as a power of the negative Laplacian, i.e., $D^{\alpha} = (-\Delta)^{\alpha/2}$. The case of negative $\text{Re } \alpha$ corresponds to Riesz potentials [66]. Since multiplication by $|y|^{\alpha}$ does not preserve the space $S(^N)$, definition (3.19) is not extendable to arbitrary $S'(^N)$- distributions.

To overcome this difficulty, Semyanistyi [63] formulated the brilliant idea to introduce another class of distributions as follows. Let $\Psi = \Psi(^N)$ be the subspace of $S(^N)$, consisting of functions ω such that $(\partial^{\nu}\omega)(0) = 0$ for all multi-indices ν. We denote by $\Phi = \Phi(^N)$ the Fourier image of Ψ, which is formed by Schwartz functions orthogonal to all polynomials. Let Φ' and Ψ' be the duals of Φ and Ψ, respectively. Two S'-distributions, that coincide in the Φ'- sense, differ from each other by a polynomial. For any Φ'-distribution g and any $\alpha \in$, the Riesz fractional derivative D^{α} g is correctly defined by the formula

$$\langle D^{\alpha}g, \omega \rangle = (2\pi)^{-N}\langle \hat{g}, |y|^{\alpha}\hat{\omega} \rangle, \quad \omega \in \Phi$$

(3.20)

Clearly, multiplication by $|y|^{\alpha}$ is a linear continuous operator on Ψ (but not on S!); see [50,60] for details and generalizations.

Lemma

Let $\alpha \notin \{0,-2,-4, \ldots\} \cup \{N, N + 2, N +4, \ldots\}$. If $f \in D_e(S^{N-1})$, then

$$E_{-\alpha} M^{1-\alpha} f = 2^{d-\alpha} D^{\alpha-d} E_{-d} M^{1-d} f$$

(3.21)

in the Φ'-sense. If, moreover, $\alpha - d = 2m$, $m = 0, 1, 2, \ldots$, and

$$(D_m f)(\theta) = 2^{-2m}\left[(-\Delta)^m E_{-d} f\right](x)|_{x=\theta}$$

(3.22)

then

$$\left(M^{1-\alpha} f\right)(\theta) = \left(D_m M^{1-d} f\right)(\theta)$$

(3.23)

pointwise for every $\vartheta \in S^{N-1}$.

Proof

Replace α by $1 - \alpha$ and by $1 - d$ in (3.17). Denoting $c_\alpha = 2^{-\alpha}\pi^{-N/2}$ and $cd = 2^{-d}\pi^{-N/2}$, we get

$$E_{-\alpha} M^{1-\alpha} f = c_\alpha [E_{\alpha-N} f]^\wedge, \qquad E_{-d} M^{1-d} f = c_d [E_{d-N} f]^\wedge$$

(in the S'-sense). Using these formulas, for any test function $w \in \Phi$ we obtain

$$\langle E_{-\alpha} M^{1-\alpha} f, \omega \rangle = c_\alpha \langle [E_{\alpha-N} f]^\wedge, \omega \rangle = c_\alpha \langle E_{\alpha-N} f, \hat{\omega} \rangle$$
$$= c_\alpha \langle E_{d-N} f, |y|^{\alpha-d} \hat{\omega} \rangle = c_\alpha \langle E_{d-N} f, [D^{\alpha-d} \omega]^\wedge \rangle$$
$$= c_\alpha \langle [E_{d-N} f]^\wedge, D^{\alpha-d} \omega \rangle = c_\alpha c_d^{-1} \langle E_{-d} M^{1-d} f, D^{\alpha-d} \omega \rangle$$
$$= 2^{d-\alpha} \langle D^{\alpha-d} E_{-d} M^{1-d} f, \omega \rangle.$$

Let now $\alpha - d = 2m$. Then $D^{\alpha-d} = (-\Delta)^m$, and the same reasoning is applicable for any C^∞- function supported in a neighborhood of the unit sphere. Hence, (3.21) holds pointwise in this specific case, and (3.23) follows.

Equalities (3.12) and (3.23) imply the following

Corollary

Let $S_K(\vartheta)$, $\vartheta \in S^{N-1}$, be a section function (3.13) of a G-invariant infinitely smooth body K in ^{N}; N = dn, n>1, d $\in$ {1, 2, 4, 8}. Let D_m be a differential operator (3.22), where

$$2m \neq N - d, N - d + 2, N - d + 4, \ldots.$$

Then

$$(D_m S_K)(\theta) = c\left(M^{1-d-2m} \rho_K^{N-d}\right)(\theta), \quad c = \frac{\pi^{N/2-d} \sigma_{d-1}}{N - d}$$

$$(3.24)$$

Intersection Bodies

We recall that ^{N} denotes the set of all origin-symmetric star bodies in $\mathbb{R}^N$. According to Lutwak [40], a body K $\in$ ^{N} is called an intersection body of a body L $\in$ ^{N} if $\rho_K(\vartheta) = \mathrm{vol}_{N-1}(L \cap \vartheta^\perp)$ for every $\vartheta \in S^{N-1}$. A wider class of intersection bodies, which is the closure of the Lutwak's class in the radial metric, was introduced by Goodey, Lutwak, and Weil [20] as a collection of bodies K $\in$ KN with the property $\rho_K = M_\mu$, where M is the Minkowski–Funk transform (3.2) and μ is an even nonnegative finite Borel measure on S^{N-1}. The class of all such measures will be denoted by $\,_e+(S^{N-1})$.

There exist several generalizations of the concept of intersection body [33,41,56,59,72]. One of them relies on the fact that the Minkowski–Funk transform M is a member of the analytic family M^α of the cosine transforms.

Definition

(See [56, Definition 5.1].) For $0 < \lambda < N$, a body $K \in\,^N$ is called a λ- intersection body if there is a measure $\mu \in\,_e +(S^{N-1})$ such that $\lambda_k^p = M^{1-\lambda}\mu$ (by Lemma 3.1, this is equivalent to $M^{1+\lambda-N}\lambda_k^p \in\,_e +(S^{N-1})$). We denote by I_λ^N the set of all such bodies.

The equality $\lambda_k^p = M^{1-\lambda}\mu$ means that for any $\varphi \in \mathcal{D}\,\mathcal{D}(S^{N-1})$,

$$\int_{S^{N-1}} \rho_K^k(\theta)\varphi(\theta)\,d\theta = \int_{S^{N-1}} \left(M^{1-\lambda}\varphi\right)(\theta)\,d\mu(\theta)$$

where for $\lambda \geq 1$, $(M^{1-\lambda}\varphi)(\vartheta)$ is understood in the sense of analytic continuation.3 If $\lambda = k$ is an integer, the class I_λ^N coincides with Koldobsky's class of k-intersection bodies and agrees with his concept of isometric embedding of the space $(\mathbb{R}^N, \|.\|_K)$ into $L-_{p'}$, $p = \lambda$ [33]. In the framework of this concept, all bodies $K \in I_\lambda^N$ can be regarded as "unit balls of N-dimensional subspaces of $L-_\lambda$".

The following statement is a consequence of the trace theorem for cosine transforms; see [56, Theorem 5.13].

Theorem

Let $1<m<N$, $\eta \in Gr_m\left(\mathbb{R}^N\right)$, and let $0 < \lambda < m$. If $K \in \mathcal{I}_\lambda^N$ in $\mathbb{R}^N$, then $K \cap \eta \in \mathcal{I}_\lambda^m$ in η.

This fact was used (without proof) in [34, Theorem 4]. In the case, when $\lambda = k$ is an integer, it was established by Milman [41]; see [56, Section 1.1] for the discussion of this statement.

WEIGHTED SECTION FUNCTIONS

Let K be an origin-symmetric convex body in $\mathbb{R}^N$. Given a point $z \in$ int(K) (the interior of K), we define the shifted radial function of K with respect to z,

$$\rho(z,v) = \sup\{\lambda > 0: z + \lambda v \in K\}, \quad (z,v) \in \Omega = int(K) \times S^{N-1} \tag{4.1}$$

which is a distance from z to the boundary of K in the direction v.

Lemma: 4.1

(See [59, Lemma 3.1].) If an origin-symmetric convex body K in $\mathbb{R}^N$ has C^m boundary ∂K, $1 \leq m \leq \infty$, then $\rho(z,v) \in C^m(\Omega)$.

Proof

We recall the proof. Consider the function

$$v = g(z,x) = \frac{x - z}{|x - z|}, \quad z \in int(K), \ x \in \partial K$$

Since ∂K is C^m, $g(z,x)$ is a Cm function in int(K) × ∂K. When z is fixed, $g(z,\cdot)$ is a C^m diffeomorphism from ∂K to S^{N-1}. By the implicit function theorem, $x = f(x,v)$ is a Cm function on Ω. Thus, $\rho(z,v) = |x-z| = |f(z,v) - z|$ is a C^m function on Ω.

It was discovered by Gardner [13] and Zhang [73], that positive solution to the Busemann– Petty problem for convex bodies K in $\mathbb{R}^3$
and $\mathbb{R}^4$ is intimately connected with the volume of parallel hyperplane
sections of those bodies; see also [33,35]. This volume, which is a hyperplane Radon transform of the characteristic function $\chi K(x)$ of K, is
represented as $A_{H,\vartheta}(t) = \text{vol}_{N-1}(K \cap \{H + t\vartheta\})$, where $t \in \mathbb{R}$, $\vartheta \in S^{N-1}$,
and H is a hyperplane through the origin perpendicular to ϑ . It was
noted in [53] and in [59, p. 492], that further progress can be achieved
if we replace $A_{H,\vartheta}(t)$ by the mean value of the i-plane Radon trans-
form [26,55] of some weighted function $f(\mathbf{x}) = |\mathbf{x}|^\beta \chi_K(\mathbf{x})$. This mean
value should be taken over all i-planes parallel to a fixed subspace ξ

$\in Gr_i(\mathbb{R}^N)$ at distance $|t|$ from the origin. Such averages for arbitrary
f (see[55, Definition 2.7]) play an important role in the theory of i
-plane Radon transforms. Similar "weighted" section functions were
later used in [36,74].

Let us proceed with a precise definition. Given a convex body $K \in \mathcal{K}^N$
, we define the weighted section function

$$A_{i,\beta}(t,\xi) = \int_{S^{N-1} \cap \xi^\perp} A_\beta(\xi + tu)\,du, \quad \xi \in Gr_i(\mathbb{R}^N), \; t \in \mathbb{R}$$

(4.2)

where

$$A_\beta(\xi + tu) = \int_{K \cap (\xi + tu)} |x|^\beta \, dx$$

(4.3)

is the i-plane Radon transform mentioned above. Clearly, $A_{i,\beta}(t, \xi)$ is an
even function of t. Let B = {x: $|x| \le 1$} be the unit ball in $\mathbb{R}^N$ and let $r_K =$
sup{t > 0: tB $\subset$ K} be the radius of the inscribed ball in K.

Lemma: 4.2

If a convex body $K \in \mathcal{K}^N$ is infinitely smooth and $\beta > m - i$, then all derivatives

$$A_{i,\beta}^{(j)}(t,\xi) = \left(\frac{d}{dt}\right)^j A_{i,\beta}(t,\xi), \quad 0 \leqslant j \leqslant m$$

are continuous in $(-_{rK'} {}_{rK}) \times Gr_i\left(\mathbb{R}^N\right)$.

Proof

Passing to polar coordinates in the plane $\xi + tu$, we get

$$A_\beta(\xi + tu) = \int_{S^{N-1} \cap \xi} a_{u,v}^\beta(t)\,dv, \quad a_{u,v}^\beta(t) = \int_0^{\rho(tu,v)} r^{i-1}(r^2 + t^2)^{\beta/2}\,dr$$

$$(4.4)$$

where $\rho(t\,u,v)$ is the radial function (4.1). It suffices to show that for $\beta > m - i$, all derivatives $(d/dt)^j$ a $a_{u,v}^\beta(t)$ $j = 0, 1, \ldots, m$, are continuous on $(-r_K, r_K)$ uniformly in $(u,v) \in (S^{N-1} \cap \xi^\perp) \times (S^{N-1} \cap \xi)$. Let, for short, $\rho(t) \equiv \rho(t\,u,v)$. If $m = 0$ and $\beta > -i$ the uniform (in u and v) continuity of $a_{u,v}^\beta(t)$ follows from Lemma 4.1. In the case $m = 1$ we have

$$(d/dt)a_{u,v}^\beta(t) = a_1(t) + a_2(t)$$

where $a1(t) = \rho^{i-1}(\rho^2 + t^2)^{\beta/2}\,d\rho/dt$ is nice and $a_2(t) = \beta t\, a_{u,v}^{\beta-2}(t)$. If $\beta > 2 - i$ we are done. Otherwise, if $1 - i < \beta \leq 2 - i$, then

$$a_2(t) = \beta t^{i+\beta-1} \int_0^{\rho/t} s^{i-1}(1 + s^2)^{\beta/2-1}\,ds \to 0, \quad \text{as } t \to 0$$

$$(4.5)$$

and the result is still true. Continuing this process, we obtain the required result for all m.

The next lemma is a slight generalization of the corresponding state-ments in [36] and [74].

Lemma: 4.3

Let K be an infinitely smooth origin-symmetric convex body in $\mathbb{R}^N$, $\xi \in$ $Gr_i(\mathbb{R}^N)$, $1 < i < N$. If $-i < \beta \leq 0$, then $A_{i,\beta}(t, \xi) \leq A_{i,\beta}(0, \xi)$. If $2 - i < \beta \leq 0$, then $(d^2/dt^2)A_{i,\beta}(t, \xi)|_{t=0} \leq 0$.

Proof

Replace $|x|^\beta$ in (4.3) by $-\beta \int_0^{1|X|} Z^{-\beta-1}\, dz$, $\beta < 0$, and change the order of integration. This gives

$$A_\beta(\xi + tu) = -\beta \int_0^\infty z^{-\beta-1}\, \mathrm{vol}_i\big((B_{1/z} \cap K) \cap (\xi + tu)\big)\, dz$$

where $B_{1/z}$ is a ball of radius $1/z$ centered at the origin. The integral on the right-hand side is well defined if $-i < \beta < 0$. Applying Brunn's theorem to the convex body $B_{1/z} \cap K$, we obtain

$$\mathrm{vol}_i\big((B_{1/z} \cap K) \cap (\xi + tu)\big) \leqslant \mathrm{vol}_i\big((B_{1/z} \cap K) \cap \xi\big)$$

which gives the first statement of the lemma. If $2-i < \beta < 0$, then, by Lemma 4.2, the derivative $(d^2/dt^2)\, A_{i,\beta}(t,\xi)$ is continuous in a neighbor-hood of $t = 0$ and the second statement of the lemma follows from the first one. In the case $\beta = 0$ the result follows if we apply Brunn's theo-rem just to K.

We recall some facts about analytic continuation (a.c.) of integrals

$$I(\alpha) = \frac{1}{\Gamma(\alpha)} \int_0^\infty t^{\alpha-1} f(t)\, dt, \quad \mathrm{Re}\,\alpha > 0$$

(4.6)

Lemma: 4.4

Let m be a nonnegative integer and take $f \in L^1 (\mathbb{R})$

- If, moreover, f is m times continuously differentiable in a neighborhood of t = 0, then $I(\alpha)$ extends analytically to $\text{Re}\alpha > -m$. In particular, for $-m < \text{Re}\alpha < -m+1$,

$$a.c.\ I(\alpha) = \frac{1}{\Gamma(\alpha)} \int_0^\infty t^{\alpha-1} \left[f(t) - \sum_{j=0}^{m-1} \frac{t^j}{j!} f^{(j)}(0) \right] dt$$

(4.7)

and

$$\lim_{\alpha \to -m} I(\alpha) = (-1)^m f^{(m)}(0)$$

(4.8)

- If m is odd and f is an even function, which is m + 1 times continuously differentiable in a neighborhood of t = 0, then (4.7) holds for $-m-1 < \text{Re}\alpha < -m+1$.

Proof. All statements are well known [17]. For instance, (ii) follows from the fact that all derivatives $f^{(j)}(t)$ of odd order are zero at t = 0 and therefore, for m odd, the sum $\Sigma_{j=0}^{m-1}$ can be replaced by $\Sigma_{j=0}^{m}$. However, (4.8) is usually proved for functions, which have at least m + 1 N continuous derivatives at t = 0. We show that it suffices to have only m continuous derivatives. The latter is important in our consideration. Let

$$(I^\lambda f)(t) = \frac{1}{\Gamma(\lambda)} \int_0^t f(s)(t-s)^{\lambda-1} dt$$

$$= \frac{t^\lambda}{\Gamma(\lambda)} \int_0^1 f(t\eta)(1-\eta)^{\lambda-1} d\eta, \quad \lambda > 0$$

be the Riemann–Liouville fractional integral of f . Note that

$$f(t) - \sum_{j=0}^{m-1} \frac{t^j}{j!} f^{(j)}(0) = \left(I^m f^{(m)} \right)(t)$$

and $t^{-m}(I^m f^{(m)})(t) \to f^{(m)}(0)/m!$ as $t \to 0$. Hence, for any $\varepsilon > 0$ there exists $\delta = \delta(\varepsilon) > 0$ such that

$$\left| t^{-m} \left(I^m f^{(m)} \right)(t) - f^{(m)}(0)/m! \right| < \varepsilon, \quad \forall t \in (0, \delta)$$

Setting $\alpha = \alpha_0 - m$, $\alpha_0 \in (0, 1)$, we obtain

$$\frac{1}{\Gamma(\alpha)} \int_0^\infty t^{\alpha-1} \left[f(t) - \sum_{j=0}^{m-1} \frac{t^j}{j!} f^{(j)}(0) \right] dt - (-1)^m f^{(m)}(0)$$

$$= \frac{1}{\Gamma(\alpha_0 - m)} \int_0^\delta t^{\alpha_0-1} \left[t^{-m} \left(I^m f^{(m)} \right)(t) - f^{(m)}(0)/m! \right] dt$$

$$+ f^{(m)}(0) \left[\frac{\delta^{\alpha_0}}{\alpha_0 \Gamma(\alpha_0 - m)m!} - (-1)^m \right]$$

$$+ \frac{1}{\Gamma(\alpha_0 - m)} \int_\delta^\infty t^{\alpha_0-m-1} \left[f(t) - \sum_{j=0}^{m-1} \frac{t^j}{j!} f^{(j)}(0) \right] dt = I_1 + I_2 + I_3$$

If $\alpha_0 \to 0$, then $\alpha_0 \Gamma(\alpha_0 - m) m! \to (-1)m$,

$$|I_1| < \frac{\varepsilon \delta^{\alpha_0}}{\alpha_0 |\Gamma(\alpha_0 - m)|} \to \varepsilon m!, \quad I_2 \to 0, \quad I_3 \to 0$$

This gives the result.

The next lemma establishes connection between weighted section functions, spherical Radon transforms, and cosine transforms.

Lemma: 4.5

Let $\xi \in \mathrm{Gr}_i \left(\mathbb{R}^N \right)$, $1 < i < N$. Suppose that

$$\alpha \neq N - i, N - i + 2, N - i + 4, \ldots.$$

and K is an infinitely smooth origin-symmetric convex body in $\mathbb{R}^N$.

- If $\beta > -i$ and $\mathrm{Re}\,\alpha > 0$, then

$$\frac{1}{\Gamma(\alpha/2)} \int_0^\infty t^{\alpha-1} A_{i,\beta}(t,\xi)\,dt = c\left(R_{N-i} M^{\alpha+1+i-N} \rho_K^{\alpha+\beta+i}\right)(\xi^\perp)$$

$$c = \frac{\pi^{i/2} \sigma_{N-i-1}}{(\alpha + \beta + i)\sigma_{N-1} \Gamma((N - i - \alpha)/2)}$$

(4.9)

- If $\beta > 1 - i$, then (4.9) extends to $-1 < \mathrm{Re}\,\alpha < 0$ as

$$\frac{1}{\Gamma(\alpha/2)} \int_0^\infty t^{\alpha-1} \left[A_{i,\beta}(t,\xi) - A_{i,\beta}(0,\xi)\right] dt = c\left(R_{N-i} M^{\alpha+1+i-N} \rho_K^{\alpha+\beta+i}\right)(\xi^\perp)$$

(4.10)

- If $\beta \_ 2 - i$, then (4.10) holds in the extended domain $-2 < \mathrm{Re}\,\alpha < 0$.
- If $\beta > m - i$ and $m \_ 0$ is even, then

$$\frac{\Gamma((1-m)/2)}{2^{m+1}\sqrt{\pi}} A_{i,\beta}^{(m)}(0,\xi) = c_1 \left(R_{N-i} M^{1-m+i-N} \rho_K^{\beta-m+i}\right)(\xi^\perp)$$

(4.11)

$$c_1 = \frac{\pi^{i/2} \sigma_{N-i-1}}{(\beta - m + i)\sigma_{N-1} \Gamma((N - i + m)/2)}$$

Proof

- Consider the integral

$$g_{\alpha,\beta}(\xi) = \frac{1}{\Gamma(\alpha/2)} \int_K |P_{\xi\perp} x|^{\alpha+i-N} |x|^\beta \, dx, \quad \mathrm{Re}\,\alpha > 0$$

(4.12)

where $P\xi_\perp$ denotes the orthogonal projection onto $\xi^\perp$. We transform (4.12) in two different ways (a similar trick was used in [54, p. 61] and [59, p. 490]). On the one hand, integration over slices parallel to ξ gives

$$g_{\alpha,\beta}(\xi) = \frac{1}{\Gamma(\alpha/2)} \int_{\xi^\perp} |y|^{\alpha+i-N}\,dy \int_{K\cap(\xi+y)} |x|^\beta\,dx$$

$$= \frac{1}{\Gamma(\alpha/2)} \int_0^\infty t^{\alpha-1} A_{i,\beta}(t,\xi)\,dt$$

(4.13)

On the other hand, passing to polar coordinates, we can express $g_{\alpha,\beta}$ as the generalized cosine transform (3.3), namely

$$g_{\alpha,\beta}(\xi) = \frac{1}{(\alpha+\beta+i)\Gamma(\alpha/2)} \int_{S^{N-1}} \rho_K(u)^{\alpha+\beta+i} |P_{\xi_\perp}u|^{\alpha+i-N}\,du = c_{\alpha,\beta}(R_i^\alpha \rho_K^{\alpha+\beta+i})(\xi)$$

$$c_{\alpha,\beta} = \frac{2\pi^{(N-1)/2}}{(\alpha+\beta+i)\sigma_{N-1}\Gamma((N-i-\alpha)/2)}.$$

Hence, by (3.9),

$$g_{\alpha,\beta}(\xi) = \frac{c_{\alpha,\beta}\sigma_{N-i-1}}{2\pi^{(N-i-1)/2}} \left(R_{N-i} M^{\alpha+1+i-N} \rho_K^{\alpha+\beta+i}\right)(\xi^\perp)$$

(4.14)

which gives (4.9).

- By Lemma 4.2 (with $m = 1$) the derivative $(d/dt)\,A_{i,\beta}(t,\xi)$ is continuous in a neighborhood of $t = 0$. Keeping in mind that

$$\lim_{\alpha\to-m} \frac{\Gamma(\alpha)}{\Gamma(\alpha/2)} = \frac{\Gamma((1-m)/2)}{2^{m+1}\sqrt{\pi}}$$

and applying Lemma 4.4(i), we obtain (4.10).

- The validity of this statement for $\beta > 2-i$ is a consequence of Lemma 4.2 (with m = 2) and Lemma 4.4(ii) (with m = 1). Consider the case $\beta = 2 - i$ which is more subtle. Denote for short $F(t) = A_{1,\beta}(t, \xi)$ and let first $\beta > 1 - i$. By Lemma 4.2 the derivative $F'(t)$ is continuous in a neighborhood of t = 0. Since F is an even function, then $F'(0) = 0$ and the left-hand side of (4.10) can be written as

$$\frac{1}{\Gamma(\alpha/2)} \int_0^\infty t^{\alpha-1} \Delta(t)\,dt, \quad \Delta(t) = F(t) - F(0) - t F'(0)$$

(4.15)

By (4.2) and (4.4),

$$\Delta(t) = \int_{S^{N-1}\cap\xi^\perp} du \int_{S^{N-1}\cap\xi} \Delta_{u,v}(t)\,dv$$

where $\nabla_{u,v}(t) = f(t) - f(0) - tf'(0)$,

$$f(t) \equiv a^\beta_{u,v}(t) = \int_0^{\rho(tu,v)} r^{i-1}(r^2 + t^2)^{\beta/2}\,dr, \quad \rho \equiv \rho(tu, v)$$

To estimate $\nabla_{u,v}(t)$, we write it as $\nabla_{u,v}(t) = I_1 + I_2$, where

$$I_1 = \int_0^{\rho(tu,v)} r^{i-1}\left[(r^2 + t^2)^{\beta/2} - r^\beta\right]dr$$

$$I_2 = \int_0^{\rho(tu,v)} r^{i+\beta-1}\,dr - \int_0^{\rho(0,v)} r^{i+\beta-1}\,dr - t\left[a_1(0) + a_2(0)\right]$$

$$a_1(t) = \rho^{i-1}\left(\rho^2 + t^2\right)^{\beta/2} d\rho/dt, \qquad \rho \equiv \rho(tu, v), \qquad a_2(t) = \beta t a_{u,v}^{\beta-2}(t)$$

For I_1, changing the order of integration, we have

$$I_1 = \frac{\beta}{2} \int_0^{\rho} r^{i-1} dr \int_0^{t^2} (r^2 + s)^{\beta/2-1} ds = \frac{\beta}{4} \int_0^{t^2} s^{(i+\beta)/2-1} h(s) ds$$

$$h(s) = \int_0^{\rho^2/s} \eta^{i/2-1}(\eta + 1)^{\beta/2-1} d\eta$$

If $\beta = 2 - i$ then $h(s) = O(\log(1/s))$ as $s \to 0$ and therefore, $I_1 = O(t^2 \log(1/t))$ as $t \to 0$. To estimate I_2 we note that $a_2(0) = 0$ (see (4.5)) and therefore,

$$I_2 = \frac{1}{i+\beta}\left[\rho(tu, v)^{i+\beta} - \rho(0, v)^{i+\beta} - t(i + \beta)\rho(0, v)^{i+\beta-1}\rho'(0, v)\right]$$

$$= \psi(t) - \psi(0) - t\psi'(0), \qquad \psi(t) \equiv \rho(tu, v)^{i+\beta}$$

Hence, $I_2 = O(t^2)$ as $t \to 0$. Since all estimates above are uniform in u and v, then the function $\nabla(t)$ in (4.15) is $O(t^2 \log(1/t))$ as $t \to 0$. This enables us to extend this integral by analyticity to all Re$\alpha > -2$.

The statement (iv) follows from Lemma 4.2 (with m = 2) and (4.8)

COMPARISON OF VOLUMES. PROOFS OF THE MAIN RESULTS

We recall basic notation from Section 2.5 related to the lower dimensional slice comparison problem (Problem B). Let K and L be origin-symmetric convex bodies in $\mathbb{R}^N$, N = dn, where n>1, d $\in$ {1, 2, 4, 8}; G is the class (2.48) of block diagonal orthogonal transformations of

$\mathbb{R}^N$, which includes the groups $G_{\mathbb{R}}, G_{\mathbb{C}}, G_{\mathbb{H},l}, G_{\mathbb{H},r}$; see (2.32)–(2.34).

The notation $\widetilde{Gr}_{N-d}(\mathbb{R}^N)$ is used for the respective manifolds (2.51) of $(N - d)$-dimensional subspaces H_ϑ $\vartheta \in S^{N-1}$, in particular, for

$$Gr_{n-1}(\mathbb{R}^n), \quad Gr_{2n-2}^{\mathbb{C}}(\mathbb{R}^{2n}), \quad Gr_{4n-4}^{\mathbb{H},l}(\mathbb{R}^{4n}), \quad Gr_{4n-4}^{\mathbb{H},r}(\mathbb{R}^{4n});$$

see Section 2.5. If K is an infinitely smooth G-invariant star body in $\mathbb{R}^N$, then, by Lemma 3.4 and Corollary 3.7,

$$S_K(\theta) \equiv \mathrm{vol}_{N-d}(K \cap H_\theta) = c\left(M^{1-d}\rho_K^{N-d}\right)(\theta) \tag{5.1}$$

$$(D_m S_K)(\theta) = c\left(M^{1-d-2m}\rho_K^{N-d}\right)(\theta), \tag{5.2}$$

where

$$c = \pi^{N/2-d}\sigma_{d-1}/(N-d), \qquad (D_m f)(\theta) = 2^{-2m}\left[(-\Delta)^m E_{-d} f\right](x)|_{x=\theta}$$

$$2m \neq N - d, N - d + 2, N - d + 4, \ldots. \tag{5.3}$$

Lemma: 5.1

Let

$$\alpha \notin \{0, -2, -4, \ldots\} \cup \{N, N+2, N+4, \ldots\}. \tag{5.4}$$

- If K, L are infinitely smooth G-invariant star bodies in $\mathbb{R}^N$ such that $\left(M^{\alpha+1-N}\rho_K^d\right)(\vartheta) \geq 0$ and

$$(M^{1-\alpha}\rho_K^{N-d})(\theta) \leqslant (M^{1-\alpha}\rho_L^{N-d})(\theta), \quad \forall \theta \in S^{N-1} \tag{5.5}$$

then $\mathrm{vol}_N(K) \le \mathrm{vol}_N(L)$.

- If L is an infinitely smooth G-invariant convex body with positive curvature such that $\left(M^{\alpha+1-N}\rho_L^d\right)(\theta) < 0$ for some $\vartheta \in S^{N-1}$, then there exists a G-invariant smooth convex body K for which (5.5) holds, but $\mathrm{vol}_N(K) > \mathrm{vol}_N(L)$.

Proof

- By Lemma 3.1,

$$N\,\mathrm{vol}_N(K) = \int_{S^{N-1}} \rho_K^N(\theta)\,d\theta = \left(\rho_K^{N-d},\rho_K^d\right) = \left(M^{1-\alpha}\rho_K^{N-d}, M^{\alpha+1-N}\rho_K^d\right)$$

Since $M^{\alpha+1-N}\rho_K^d > 0$, we can continue:

$$N\,\mathrm{vol}_N(K) \le \left(M^{1-\alpha}\rho_L^{N-d}, M^{\alpha+1-N}\rho_K^d\right) = \left(\rho_L^{N-d}, \rho_K^d\right)$$

Now the result follows by Hölder's inequality.

- Let $\varphi(\vartheta) \equiv \left(M^{\alpha+1-N}\rho_L^d\right)(\theta) < 0$ for some $\vartheta \in S^{N-1}$. Then φ is negative on some open set $\Omega \subset S^{N-1}$ and, by Lemma 3.1, $\rho_L^d = M^{1-\alpha}\varphi$. Since φ is G-invariant, then $\varphi < 0$ on the whole orbit $G\Omega$. Choose a function $\psi \in \mathcal{D}(S^{N-1})$ so that $\psi \ne 0$, $\psi(\vartheta) > 0$ if $\vartheta \in G\Omega$, and $\psi(\vartheta) \equiv 0$ otherwise. Without loss of generality, we can assume ψ to be G-invariant (otherwise, it can be replaced by $\tilde{\psi}(\theta) = \int_G \psi(\gamma\theta)\,d\gamma$).

  Define a smooth G-invariant body K by $\rho_K^{N-d} = \rho_L^{N-d}\varepsilon M^{\alpha+1-N}\psi, \varepsilon > 0$. If ε is small enough, then K is convex. This conclusion is a consequence of Oliker's formula [43], according to which the Gaussian curvature of an origin-symmetric star body expresses through the first and second derivatives of the radial function. Applying $M^{1-\alpha}$ to the preceding equality, we obtain

$$M^{1-\alpha}\rho_K^{N-d} - M^{1-\alpha}\rho_L^{N-d} = -\varepsilon M^{1-\alpha} M^{\alpha+1-N}\psi = -\varepsilon\psi \leqslant 0$$

which gives (5.5). On the other hand

$$\left(\rho_L^d, \rho_L^{N-d} - \rho_K^{N-d}\right) = \varepsilon\left(M^{1-\alpha}\varphi, M^{\alpha+1-N}\psi\right) = \varepsilon(\varphi, \psi) < 0$$

or $\left(\rho_L^d, \rho_L^{N-d}\right) < \left(\rho_L^d, \rho_K^{N-d}\right)$. By Hölder's inequality, the latter implies $vol_N(L) < vol_N(K)$.

Now, we investigate for which α the inequality $\left(M^{\alpha+1-N}\rho_K^d\right)(\theta) \geq 0$ in Lemma 5.1 is available.

Lemma: 5.2

Let K and L be infinitely smooth G-invariant convex bodies in $\mathbb{R}^N$; $N = dn$, $n>1$, $d \in \{1, 2, 4, 8\}$. Suppose that

$$\left(M^{1-\alpha}\rho_K^{N-d}\right)(\theta) \leqslant \left(M^{1-\alpha}\rho_L^{N-d}\right)(\theta), \quad \forall \theta \in S^{N-1}$$

for some α satisfying

$$\max(N - d - 2, d) \leqslant \alpha < N. \tag{5.6}$$

Then $vol_N(K) \leq vol_N(L)$.

Proof

We apply Lemma 4.5 with $\xi = H_\vartheta$, $i = N - d$, and α replaced by $\alpha + d - N$. By Lemma 2.14 the expression $(R_{N-i}M^{\alpha+1+i-N}\rho_K^{\alpha+\beta+i})$ $(\xi^\perp)$ in Lemma 4.5 transforms into $I\alpha, \beta = \left(M^{\alpha+1-N}\rho_K^{\alpha+\beta}\right)(\theta)$ and the latter is represented as follows.

- For $\alpha > N - d$, $\beta > d - N$:

$$I_{\alpha,\beta} = \frac{c^{-1}}{\Gamma((\alpha+d-N)/2)} \int_0^\infty t^{\alpha+d-N-1} A_{N-d,\beta}(t, H_\theta) \, dt$$

(5.7)

- For $\alpha = N - d$, $\beta > d - N$:

$$I_{\alpha,\beta} = \frac{1}{2} A_{N-d,\beta}(0, H_\theta)$$

(5.8)

- For (a) $N - d - 1 < \alpha < N - d$, $1 + d - N < \beta \le 0$, and (b) $N - d - 2 < \alpha < N - d$, $2 + d - N \le \beta \le 0$:

$$I_{\alpha,\beta} = \frac{c^{-1}}{\Gamma((\alpha+d-N)/2)} \int_0^\infty t^{\alpha+d-N-1} \left[A_{N-d,\beta}(t, H_\theta) - A_{N-d,\beta}(0, H_\theta) \right] dt$$

(5.9)

- For $\alpha = N - d - 2$, $2 + d - N < \beta \le 0$:

$$I_{\alpha,\beta} = -\frac{c_1^{-1}}{4} A''_{N-d,\beta}(0, H_\theta)$$

(5.10)

Owing to Lemma 4.3, expressions (5.7)–(5.10) are nonnegative. Set $\beta = d - \alpha$ to get $M^{\alpha+1-N} \rho_K^d \equiv I_{\alpha, d-\alpha}$. Then combine inequalities in each case. We obtain the following bounds for α.

For $d = 1$, $N = n$: $\max(n - 3, 1) \le \alpha < n$.

For $d = 2, 4, 8$:

(5.7) holds if $N - d < \alpha < N$;

(5.8) holds if $\alpha = N - d$;

(5.9) holds if $N - d - 1 < \alpha < N - d$ when $N \ge 2d + 1$;

$\qquad\qquad N - d - 2 \le \alpha < N - d$ when $N \ge 2d + 2$;

$\qquad\qquad d \le \alpha < N - d$ when $2d < N < 2d + 2$;

(5.10) holds if $\alpha = N - d - 2$, $N \geq 2d + 2$.

Combining these inequalities, we obtain (5.6).

Remark: 5.3

The operator $M^{1-\alpha} \equiv (M^{1+\alpha-N})^{-1}$ in Lemmas 5.1 and 5.2, that was originally defined by analytic continuation of the integral (3.5), can be explicitly represented as an integrodifferential operator $P(\delta) M^{\nu}$, where M^{ν}, $\nu > 0$, has the form (3.5) and $P(\delta)$ is a polynomial of the Beltrami–Laplace operator δ on S^{N-1}; see [51, Section 2.2] for details.

Lemma 5.2 leads to main results of the paper. The next statement gives a positive answer to Problem B.

Theorem: 5.4

Let K and L be G-invariant convex bodies in $\mathbb{R}^N$ with section functions

$$S_K(\theta) = \mathrm{vol}_{N-d}(K \cap H_\theta), \qquad S_L(\theta) = \mathrm{vol}_{N-d}(L \cap H_\theta),$$

where $H_\theta \in \widetilde{\mathrm{Gr}}_{N-d}(\mathbb{R}^N)$, N = dn, n>1, d $\in$ {1, 2, 4, 8}. Suppose that

$$S_K(\theta) \leqslant S_L(\theta), \quad \forall \theta \in S^{N-1}. \tag{5.11}$$

If n≤2+ 2/d, then vol_N (K) ≤ vol_N (L).

Proof

For infinitely smooth bodies the result is contained in Lemma 5.2 (set $\alpha = d$ and make use of (5.1)). Let us extend this result to arbitrary G-invariant convex bodies. Given a G-invariant convex body K, let

$$K^* = \{x : |x \cdot y| \leqslant 1, \forall y \in K\}$$

be the polar body of K with support function

$$h_{K*}(x) = \max\{x \cdot y : y \in K^*\}.$$

Since $h_K * (\cdot)$ coincides with Minkowski's functional $\|.\|K$, then $h_{K*}(\cdot)$ is G-invariant, and therefore, $K*$ is G-invariant too. It is known [62, pp. 158–161] that any origin-symmetric convex body in $\mathbb{R}^N$ can be approximated by infinitely smooth convex bodies with positive curvature and the approximating operator commutes with rigid motions. Hence, there is a sequence $\{K_j^*\}$ of infinitely smooth G-invariant convex bodies with positive curvature such that $h_{K_j^*}(\theta)$ converges to uniformly on S^{N-1}. The latter means, that for the relevant sequence of infinitely smooth G-invariant convex bodies $K_j = (K_j^*)^*$ we have

$$\lim_{j \to \infty} \max_{\theta \in S^{N-1}} \left| \|\theta\|_{K_j} - \|\theta\|_K \right| = 0.$$

This implies convergence in the radial metric, i.e.,

$$\lim_{j \to \infty} \max_{\theta \in S^{N-1}} \left| \rho_{K_j}(\theta) - \rho_K(\theta) \right| = 0. \tag{5.12}$$

Let us show that the sequence $\{K_j\}$ in (5.12) can be modified so that $K_j \subset K$. An idea of the

argument was borrowed from [59]. Without loss of generality, assume that $\rho_K(\vartheta) \geq 1$. Choose K_j so that

$$\left| \rho_{K_j}(\theta) - \rho_K(\theta) \right| < \frac{1}{j+1}, \quad \forall \theta \in S^{N-1},$$

and set $K_j' = \dfrac{j}{j+1} K_j$. Then, obviously, $\rho_{K_j'}(\theta) \to \rho_K(\theta)$ uniformly on S^{N-1} as $j \to \infty$, and

$$\rho_{K_j'} = \frac{j}{j+1}\rho_{K_j} < \frac{j}{j+1}\left(\rho_K + \frac{1}{j+1}\right) \leqslant \rho_K .$$

Hence, $K_j' \subset K$. Now suppose that (5.11) is true. Then it is true when K is replaced by K_j', and, by the assumption of the lemma, $\mathrm{vol}_N (K_j') \leq \mathrm{vol}_N$ (L). Passing to the limit as $j \to \infty$, we obtain vol_N (K) $\leq \mathrm{vol}_N$ (L).

The following theorem, which generalizes Theorem 4 from [34], shows that the restriction $n \leq 2 + 2/d$ in Theorem 5.4 is sharp.

Theorem: 5.5

Let $N = dn > 2d+2, n > 1$, $d \in \{1, 2, 4, 8\}$. Then there exist G-invariant infinitely smooth convex bodies K and L in $\mathbb{R}^N$ such that $S_K (\vartheta) \leq S_L(\vartheta)$ for all $\vartheta \in S^{N-1}$, but $\mathrm{vol}_N(K) > \mathrm{vol}N(L)$.

Proof

Let $x = (x_1, \ldots, x_n)^T \in \mathbb{R}^N$, $x_j = (x_{j,1} \ldots, x_{j,d})^T$,

$$L = \left\{ x: \|x\|_4 = \left(\sum_{j=1}^{n}|x_j|^4\right)^{1/4} \leqslant 1 \right\}.$$

Clearly, L is a G-invariant infinitely smooth convex body. Let X be the $(N - d + 1)$-dimensional subspace of $\mathbb{R}^N$, which consists of vectors of the form $(x_{1,1}, x_2, \ldots, x_n)T$. By [33, Theorems 4.19, 4.21], $L \cap X$ is not a λ-intersection body in $\mathbb{R}^{N-d+1}$ if $0 < \lambda < N - d - 2$. Hence, by Theorem 3.9, L is not a λ-intersection body for such λ. It means (see Definition 3.8) that $\left(M^{1+\lambda-N}\rho_K^\lambda\right)(\theta) < 0$ for some $\vartheta \in S^{N-1}$. Set $\lambda = d$ to get $dn > 2d + 2$ and apply Lemma 5.1(ii) with $\alpha = d$. This gives the result.

Corollary: 5.6

The hyperplane slice comparison problem (Problem A) in ^{n}, $n > 1$, has an affirmative answer if and only if $n \leq 2 + 2/d$. In particular, in $\mathbb{R}^N$: if and only if $n \leq 4$;

in $\mathbb{C}^N$: if and only if $n \leq 3$;

in $\mathbb{H}_l^n$ and $\mathbb{H}_r^n$: if and only if $n = 2$.

Theorem 5.4 also implies the following.

Corollary: 5.7

Let G be the subgroup of orthogonal transformations of $\mathbb{R}^N$ defined by (2.48)–(2.47), and let $d \in \{2, 4, 8\}$, $i = N - d$. The lower dimensional slice comparison problem (Problem B) for i-dimensional sections of N-dimensional G-invariant convex bodies has an affirmative answer in the following cases:

- $N = 4$ ($d = 2$): $i = 2$,

- $N = 6$ ($d = 2$): $i = 4$,

- $N = 8$ ($d = 4$): $i = 4$,

- $N = 10$ ($d = 4$): $i = 6$,

- $N = 16$ ($d = 8$): $i = 8$.

To make this statement a little bit more transparent to the reader, we recall, say, for $d = 8$, that the group G is formed by orthogonal transformations of $\mathbb{R}^{16}$ of the form

$$g = \begin{bmatrix} \sum_{i=0}^{7} \lambda_i A_i & 0 \\ 0 & \sum_{i=0}^{7} \lambda_i A_i \end{bmatrix} \quad \text{(2 equal blocks)}$$

where $\lambda = (\lambda_0, \lambda_1, \ldots, \lambda_7)$ is a point of S^7, and $\{A_i\}_{i=1}^{7}$ is a complete system of orthonormal tangent linear vector fields on S^7, which is assumed to be fixed; see Example 2.10

Another consequence of Lemma 5.2, which addresses Problem C, can be obtained if we set

$\alpha = d + 2m$ in that lemma and make use of Corollary 3.7.

Theorem 5.8

Let K and L be infinitely smooth G-invariant convex bodies in $\mathbb{R}^N$; N = dn, n>1, d $\in$ {1, 2, 4, 8}. Suppose that

$$(-\Delta)^m E_{-d} S_K(\theta) \leqslant (-\Delta)^m E_{-d} S_L(\theta), \quad \forall \theta \in S^{N-1},$$

for some m satisfying max

$$\max(N - 2d - 2, 0) \leqslant 2m < N - d. \tag{5.13}$$

Then $vol_N (K) \leq vol_N (L)$. In particular, m can be chosen as follows.

For d = 1: m = 0 if n $\leq$ 4, and m $\in \left\{\dfrac{n-4}{2}, \dfrac{n-3}{2}, \dfrac{n-2}{2}\right\}$ if n>4.

For d = 2: m = 0 if n $\leq$ 3, and m $\in$ {n− 3, n−2} if n>3.

For d = 4: m = 0 if n = 2, and m $\in$ {2n−5, 2n− 4, 2n− 3} if n>2.

For d = 8: m = 0 if n = 2, and m $\in$ {4n−9, 4n−8, 4n−7, 4n−6, 4n−5} if n>2.

ACKNOWLEDGMENTS

I am grateful to Professors Ralph Howard, Daniel Sage, Irina Markina, and Michael Shapiro for useful discussions. My sincere gratitude is addressed to the referee, who carefully read the manuscript and made numerous valuable suggestions that improved the original text.

REFERENCES

1. J.F. Adams, Vector fields on spheres, Ann. of Math. 75 (1962) 603–632.
2. E. Artin, Geometric Algebra, Wiley–Interscience, 1988.
3. K. Ball, Some remarks on the geometry of convex sets, in: Geometric Aspects of Functional Analysis (1986/87), in: Lecture Notes in Math., vol. 1317, Springer-Verlag, Berlin, Heidelberg, New York, 1988, pp. 224–231.
4. F. Barthe, M. Fradelizi, B. Maurey, A short solution to the Busemann–Petty problem, Positivity 3 (1999) 95–100.
5. A. Barvinok, a Course in Convexity, Grad. Stud. Math., vol. 54, Amer. Math. Soc., Providence, RI, 2002.
6. R. Bellman, Introduction to Matrix Analysis, 2nd edition, Society for Industrial and Applied Mathematics, 1997.
7. R. Bott, J. Milnor, on the parallelizability of the spheres, Bull. Amer. Math. Soc. 64 (1958) 87–89.
8. J. Bourgain, G. Zhang, on a generalization of the Busemann–Petty problem, in: Convex Geometric Analysis, Berkeley, CA, 1996, in: Math. Sci. Res. Inst. Publ., vol. 34, Cambridge University Press, Cambridge, 1999, pp. 65–76.
9. H. Busemann, C.M. Petty, Problems on convex bodies, Math. Scand. 4 (1956) 88–94.
10. O. Calin, D.-C. Chang, I. Markina, Geometric analysis on H-type groups related to division algebras, Math. Nachr. 282 (1) (2009) 44–68.
11. H. Cuypers, Regular quaternionic polytopes, Linear Algebra Appl. 226/228 (1995) 311–329.
12. B. Eckmann, Gruppentheoretischer Beweis des Satzes von Hurwitz–Radon über die Komposition quadratischer Formen, Comment. Math. Helv. 15 (1943) 358–366.
13. R.J. Gardner, A positive answer to the Busemann–Petty problem in three dimensions, Ann. of Math. (2) 140 (1994) 435–447.
14. R.J. Gardner, Intersection bodies and the Busemann–Petty problem, Trans. Amer. Math. Soc. 342 (1994) 435–445.
15. R.J. Gardner, Geometric Tomography, 2nd edition, Cambridge University Press, New York, 2006.
16. R.J. Gardner, A. Koldobsky, T. Schlumprecht, An analytic solution to the Busemann–Petty problem on sections of convex bodies, Ann. of Math. (2) 149 (1999) 691–703.
17. I.M. Gelfand, G.E. Shilov, Generalized Functions, vol. 1: Properties and Operations, Academic Press, New York, 1964.
18. A. Giannopoulos, A note on a problem of H. Busemann and C.M. Petty concerning sections of symmetric convex bodies, Mathematika 37 (1990) 239–244.
19. I.M. Glazman, Ju.I. Ljubic, Finite-Dimensional Linear Analysis: A Systematic Presentation in Problem Form, Dover Publications, 2006.

20. P. Goodey, E. Lutwak, W. Weil, Functional analytic characterizations of classes of convex bodies, Math. Z. 222 (1996) 363–381.
21. E. Grinberg, I. Rivin, Infinitesimal aspects of the Busemann–Petty problem, Bull. Lond. Math. Soc. 22 (1990) 478–484.
22. E.L. Grinberg, G. Zhang, Convolutions, transforms, and convex bodies, Proc. Lond. Math. Soc. (3) 78 (1999) 77–115.
23. X. Gual, A.M. Naveira, A. Tarrío, Integral geometry in Euclidean and projective quaternionic spaces, Bull. Math. Soc. Sci. Math. Roumanie 43 (91) (2000) 267–277. B. Rubin / Advances in Mathematics 225 (2010) 1461–1498 1497
24. X. Gual, A.M. Naveira, A. Tarío, An introduction to integral geometry in the n-dimensional quaternionic space, Gen. Math. 5 (1995) 171–177.
25. H. Hadwiger, Radialpotenzintegrale zentralsymmetrischer rotationkörper und ungleichheitsaussagen Busemannscher art, Math. Scand. 23 (1968) 193–200.
26. S. Helgason, the Radon Transform, Birkhäuser, Boston, 1999.
27. L. Hesselholt, Vector fields on sphere, preprint, http://www-math.mit.edu/~larsh/teaching/vectorfields.pdf.
28. A.S. Householder, the Theory of Matrices in Numerical Analysis, Blaisdell Pub. Co., New York, 1964.
29. A. Hurwitz, Über die Komposition der quadratischen Formen, Math. Ann. 88 (1923) 1–25.
30. D. Husemöller, Fibre Bundles, 3rd edition, Springer, 1993.
31. V.I. Istratescu, Inner Product Structures: Theory and Applications, D. Reidel Publ. Comp., Dordrecht, 1987.
32. M.A. Kervaire, Non-parallelizability of the n-sphere for n > 7, Proc. Natl. Acad. Sci. USA 44 (1958) 280–283.
33. A. Koldobsky, Fourier analysis in Convex Geometry, Math. Surveys Monogr., vol. 116, Amer. Math. Soc., 2005.
34. A. Koldobsky, H. König, M. Zymonopoulou, The complex Busemann–Petty problem on sections of convex bodies, Adv. Math. 218 (2008) 352–367.
35. A. Koldobsky, V. Yaskin, the Interface between Convex Geometry and Harmonic Analysis, CBMS Reg. Conf. Ser. Math., vol. 108, Amer. Math. Soc., Providence, RI, 2008.
36. A. Koldobsky, V. Yaskin, M. Yaskina, Modified Busemann–Petty problem on sections of convex bodies, Israel J. Math. 154 (2006) 191–207.
37. V.V. Kravchenko, M. Shapiro, Integral Representations for Spatial Models of Mathematical Physics, Longman, 1996.
38. D.G. Larman, C.A. Rogers, The existence of a centrally symmetric convex body with central sections that are unexpectedly small, Mathematika 22 (1975) 164–175.
39. P. Lounesto, Clifford Algebras and Spinors, Cambridge University Press, Cambridge, 2001.
40. E. Lutwak, Intersection bodies and dual mixed volumes, Adv. Math. 71 (1988) 232–261.

41. E. Milman, Generalized intersection bodies, J. Funct. Anal. 240 (2006) 530–567.
42. A.A. Ognikyan, Combinatorial construction of tangent vector fields on spheres, Math. Notes 83 (2008) 539–553.
43. V.I. Oliker, Hypersurfaces in Rn+1 with prescribed Gaussian curvature and related equations of Monge–Ampère type, Comm. Partial Differential Equations 9 (1984) 807–838.
44. M. Papadimitrakis, On the Busemann–Petty problem about convex centrally symmetric bodies in Rn, Mathematika 39 (1992) 258–266.
45. I.R. Porteous, Topological Geometry, 2nd edition, Cambridge University Press, Cambridge, New York, 1981.
46. I.R. Porteous, Clifford Algebras and the Classical Groups, Cambridge University Press, 1995.
47. J. Radon, Lineare Scharen orthogonaler Matrizen, Abh. Sem. Hamburg I (1923) 1–14.
48. A.P. Robertson, W.J. Robertson, Topological Vector Spaces, Cambridge Tracts in Math., vol. 53, Cambridge University Press, 1964.
49. B.A. Rosenfeld, Non-Euclidean Geometries, Gos. Izd. Tekh. Teor. Lit., Moscow, 1955 (in Russian).
50. B. Rubin, Fractional Integrals and Potentials, Addison–Wesley/Longman, Essex, UK, 1996.
51. B. Rubin, Inversion of fractional integrals related to the spherical Radon transform, J. Funct. Anal. 157 (1998) 470–487.
52. B. Rubin, Inversion formulas for the spherical Radon transform and the generalized cosine transform, Adv. in Appl. Math. 29 (2002) 471–497.
53. B. Rubin, Analytic families associated to Radon transforms in integral geometry, Lecture delivered at the PIMS, Vancouver, and July 1–5, 2002.
54. B. Rubin, Notes on Radon transforms in integral geometry, Fract. Calc. Appl. Anal. 6 (2003) 25–72.
55. B. Rubin, Reconstruction of functions from their integrals over k-planes, Israel J. Math. 141 (2004) 93–117.
56. B. Rubin, Intersection bodies and generalized cosine transforms, Adv. Math. 218 (2008) 696–727.
57. B. Rubin, The lower dimensional Busemann–Petty problem for bodies with the generalized axial symmetry, Israel J. Math. 173 (2009) 213–233.
58. B. Rubin, Comparison of volumes of convex bodies in real, complex, and quaternionic spaces, preprint, arXiv: 0812.1300v2, 2009.
59. B. Rubin, G. Zhang, Generalizations of the Busemann–Petty problem for sections of convex bodies, J. Funct. Anal. 213 (2004) 473–501.
60. S.G. Samko, A.A. Kilbas, O.I. Marichev, Fractional Integrals and Derivatives. Theory and Applications, Gordon and Breach, London, 1993.
61. L.A. Santaló, Integral Geometry and Geometric Probability, 2nd edition, Cambridge Math. Lib., Cambridge University Press, Cambridge, 2004.

62. R. Schneider, Convex Bodies: The Brunn–Minkowski Theory, Cambridge University Press, 1993.
63. V.I. Semyanistyi, on some integral transformations in Euclidean space, Dokl. Akad. Nauk SSSR 134 (1960) 536– 539 (in Russian).
64. V.I. Semyanistyi, Some integral transformations and integral geometry in an elliptic space, Tr. Sem. Vekt. Tenzor. Anal. 12 (1963) 397–441 (in Russian).
65. A.P. Shirokov, Geometry of tangent bundles and spaces over algebras, J. Math. Sci. 21 (2) (1983) 151–177.
66. E.M. Stein, Singular Integrals and Differentiability Properties of Functions, Princeton University Press, Princeton, NJ, 1970.
67. K. Tapp, Matrix Groups for Undergraduates, Stud. Math. Libr., vol. 29, Amer. Math. Soc., Providence, RI, 2005.
68. C. Truesdell, The influence of elasticity on analysis: The classic heritage, Bull. Amer. Math. Soc. 9 (1983) 293–310.
69. V.V. Vishnevskii, A.P. Shirokov, V.V. Shurygin, Spaces over Algebras, Kazanskii Gosudarstvennyi Universitet, Kazan, 1985 (in Russian).
70. Y.C. Wong, Linear Geometry in Euclidean 4-Space, Southeast Asian Math. Soc. Monogr., vol. 1, SEAMS, Hong Kong, 1977.
71. F. Zhang, Quaternions and matrices of quaternions, Linear Algebra Appl. 251 (1997) 21–57.
72. G. Zhang, Sections of convex bodies, Amer. J. Math. 118 (1996) 319–340.
73. G. Zhang, A positive solution to the Busemann–Petty problem in R4, Ann. of Math. (2) 149 (1999) 535–543.
74. M. Zymonopoulou, The modified complex Busemann–Petty problem on sections of convex bodies, arXiv: 0807.0776, 2008.

CITATION

Boris Rubin, Comparison of volumes of convex bodies in real, complex, and quaternionic spaces, Advances in Mathematics, Volume 225, Issue 3, 20 October 2010, Pages 1461-1498, ISSN 0001-8708, http://dx.doi.org/10.1016/j.aim.2010.04.005.

Constructive Mathematics:
A Foundation for Computable Analysis

Douglas S. Bridges
Department of Mathematics & Statistics,
University of Canterbury, Christchurch,
New Zealand

ABSTRACT

This paper introduces Bishop's constructive mathematics, which can be regarded as the constructive core of mathematics and whose theorems can be translated into many formal systems of computable analysis. The real numbers are presented using a set of constructive axioms, from which are derived some elementary properties of the real line R, including its completeness.

WHAT IS CONSTRUCTIVE MATHEMATICS?

In this paper we outline some of the main features of modem constructive mathematics, in the hope that we may convince our colleagues interested in all aspects of computability in analysis that, by examining the analysis developed by Errett Bishop and his followers - an analysis that is, in practice, based on logical, rather than overtly algorithmic, considerations - they can find a rich source of material for implementation within their models of computation.

Many mathematicians believe that what distinguishes constructive mathematicians from the practitioners of traditional, or classical, mathematics (that which is taught in most university courses) is a distrust of such principles as the law of excluded middle (LEM) and the axiom of choice. In fact, the original motivation for constructive mathematics, according to pioneers of the subject such as Brouwer [1 1] and Bishop [2], is a positive one: a desire to interpret existence strictly in terms of computability, or constructivity, a desire which happens to lead, for good reasons, to that distrust.

We concentrate on Errett Bishop's approach to constructive mathematics (BISH), which originated in 1967 with the publication of the book Foundations of Constructive Analysis [2], in which Bishop developed large parts of classical and modem analysis within a strictly constructive framework. His development was based on a primitive, unspecified notion of algorithm and on the properties of the natural numbers:

The primary concern of mathematics is number, and this means the positive integers. We feel about number the way Kant felt about space. The positive integers and their arithmetic are presupposed by the very nature of our intelligence and, we are tempted to believe, by the very nature of intelligence in general. The development of the positive integers from the primitive concept of the unit, the concept of adjoining a unit, and the process of mathematical induction carries complete conviction. In the words of Kronecker, the positive integers were created by God.

We have already alluded to intuitionistic logic, the logic that is forced upon us when we want to work constructively. An examination of constructive mathematical practice leads us to the following interpretations of the logical connectives and quantifiers.

- $P \vee Q$: we have either a proof of P or a proof of Q.
- $P \wedge Q$: We have a proof of P and a proof of Q.
- $\neg P$: Assuming P, we can derive a contradiction (such as $0 = 1$).
- $P \Rightarrow Q$: We can convert[2] any proof of P into a proof of Q.
- $\exists x P(x)$: There is an algorithm which computes an object x and demonstrates that P(x) holds.
- $\forall x \in A P(x)$: There is an algorithm which, applied to an object x and a proof that x E A, demonstrates that P(x) holds.

$P \vee \neg P$, whose constructive inadmissibility follows immediately from the interpretation of disjunction given earlier (a matter that we shall clarify shortly).

Now, our experience shows that when we do constructive mathematics, we are actually doing mathematics with intuitionistic logic. The desire for algorithmic interpretability forces us to use intuitionistic logic, and that restriction of our logic seems to result, inevitably, in arguments that are entirely algorithmic in character. Maybe this is Bishop's "secret still on the point of being blabbed" ([2], epigraph): algorithmic mathematics appears to be equivalent to mathematics that uses only intuitionistic logic. If that is the case - and all the evidence of our experience suggests that it is - then we can carry out our mathematics using intuitionistic logic on any reasonably defined mathematical objects, not just some class of "constructive objects". This view of constructive mathematics is essentially that of Richman [24, 25]; it was also prefigured in [4].

In order to develop mathematics, and not just logic, constructively, we need objects upon which we can operate. These objects are the natural numbers, already discussed above, together with sets and functions.

According to Bishop

A set is not an entity which has an ideal existence: a set exists only when it has been defined. To define a set we prescribe, at least implicitly, what we (the constructing intelligence) must do in order to construct an element of the set, and what we must do to show that two elements of the set are equal. [2, p. 21

Note that it is not required that the property characterizing a set be decidable (cf. [l&5]). This is consonant with recursive mathematics, in which not every set is recursive. Note also that equality is defined for each particular set, and must satisfy the usual rules of reflexivity, symmetry, and transitivity. So in order to combine sets using the operations of union and intersection, we need those sets to be given as subsets of some larger set inducing the equality on each of the subsets.

In general, Bishop does not deal with intensional equality (identity) of objects. For example, he defines a real number as a sequence (x_n) of rational numbers regular, in the sense that

$$\left|x_m - x_n\right| \leq \frac{1}{m} + \frac{1}{n}$$

for all m, n ≥ 1; he then defines two real numbers (x_n), (y_n) to be equal if

$$\left|x_m - y_n\right| \leq \frac{2}{n}$$

for all n≥ 1. So he works directly with Cauchy sequences, rather than, as would the classical mathematician, with equivalence classes of Cauchy sequences (cf. the standard practice of identifying the fractions

$\frac{1}{2}$ and $\frac{17}{34}$).

Having dealt with sets, Bishop turns to functions:

In order to define a function from a set A to a set B, we prescribe a finite routine which leads from an element of A to an element of B, and show that equal elements of A give rise to equal elements of B. [2, p. 2]

If we omit the requirement that equality be preserved, then we relax the notion of function to that of operation. This notion is rather contentious, and we will avoid further mention of it wherever possible.

It is, however, important in Martin-LÖf's type-theoretic formalization of constructive mathematics, which has had a significant impact on the work of computer scientists interested in extracting programs from proofs [21,12,17].

Building on the positive integers, weaving a web of ever more sets and more functions, we get the basic structures of mathematics: the rational number system, the real number system, the euclidean spaces, the complex number system, the algebraic number fields, Hilbert space, the classical groups, and so forth. Within the framework of these structures most mathematics is done. Everything attaches itself to number, and every mathematical statement ultimately expresses the fact that if we perform certain computations within the set of positive integers, we shall get certain results. [2, pp. 2-3].

An interesting formal system for Bishop's mathematics, using number, set, and function as primitives, was given by Myhill [23]. The same author has also developed intuitionistic Zermelo-Fraenkel set theory as a foundation for constructive mathematics [22]. Other foundational systems for BISH are found in [15, 14, 21, 7]. Excellent sources for constructive foundational matters are [1, 26].

Of course, there are many classical results that we cannot hope to prove constructively; in other words, results that are essentially no algorithmic. To illustrate this point, consider the following simple statement, dubbed the limited principle of omniscience (LPO) by Bishop:

$$\forall a \in \{0,1\}^N \ (a = 0 \vee a \neq 0)$$

where $a = (a_0, a_1, a_2, \ldots)$, $N = (0, 1, 2, \ldots)$ is the set of natural numbers,

$$a = 0 \Leftrightarrow \forall n \ (a_n = 0)$$

$$a \neq 0 \Leftrightarrow \exists n \ (a_n = 1).$$

In words, LPO states that for each binary sequence (a_n), either $a_n = 0$ for all n or else there exists n such that $a_n = 1$. If this principle could be proved constructively, then, applying it to the binary sequence defined by

$$a_n \begin{cases} 0 \text{ if } 2k \text{ is a sum of tow primes for } 2 \leq k \leq n+1 \\ 1 \text{ otherwise} \end{cases}$$

we would be able either to prove the Goldbach Conjecture - every even integer >2 is a sum of two primes - or else to compute an even integer >2 that is not a sum of two primes. By applying LPO to other well-chosen binary sequences, we would be able to solve, by a decision procedure, a whole host of open problems, including the Riemann hypothesis. "Of course, such a method . . . does not exist, and nobody expects that one will ever be found" [2, p. 51. We therefore conclude that LPO should not be admitted as a working principle of constructive mathematics, and that any proposition that constructively implies LPO should be regarded as essentially nonconstructive. In particular, this reinforces the view that the law of excluded middle, of which LPO is a trivial consequence, is no constructive.

Our rejection of LPO has some significant consequences even at the level of the real number line R. For example, we cannot expect to prove constructively that

$$\forall x \in R \ \left(x = 0 \vee x \neq 0 \right)$$

where $x \neq 0$ means $|x| > 0$. (Here we are assuming some elementary constructive properties of R.) For if we could prove this statement, then, given any binary sequence (a_n) and applying it to the real number whose binary expansion is $0. a_1 a_2 a_3 \ldots$, we could prove LPO.

Among other classical propositions that imply LPO are:

- The law of trichotomy: $\forall x \in R \ \left(x < 0 \vee x = 0 \vee x > 0 \right)$;

- The least-upper-bound principle: Each nonempty subset of R that is bounded above has a least upper bound;

- Every real number is either rational or irrational. (Consider a decreasing binary sequence (a_n) and the real number $\sum_{n=1}^{\infty} a_n / n!$).

Now, in constructive mathematics we are more interested in positive results, not these negative ones that show what cannot be done. Fortunately, we have the following constructive principles that enable us to circumvent the unavailability of the law of trichotomy and the least-upper-bound principle.

If a < b, then for each x E R either x >a or x < b. (This principle is often used to split a proof into cases).

The constructive least-upper-bound principle: If S is a nonempty subset of R that is bounded above, and if for all α, β with $\alpha < \beta$,

$$\forall x \in S(x < \beta) \vee \exists x \in S(x > \alpha)$$

then the supremum of S exists.

Another classically trivial consequence of LEM that is not acceptable in BISH is the lesser limited principle of omniscience (LLPO):

For each binary sequence (a,) with at most one term equal to 1, either $a_{2n} = 0$ for all n or else $a_{2n} + 1 = 0$ for all n.

Among the classical propositions that entail LLPO and are therefore regarded as essentially nonconstructive are

- $\forall x \in R(x \geq 0 \vee x \leq 0)$
- If $x, y \in R$ and $xy = 0$, then $x = 0$ or $y = 0$.
- The Intermediate Value Theorem: If $f : f : [0,1] \to R$ is a continuous function with $f(0) < 0 < f(1)$, then there exists $x \in (0,1)$ such that $f(x) = 0$

For more on LPO, LLPO, and related matters, we refer the reader to Ch. 1 of [10].

Again, there are constructive substitutes for these unusable classical results. For example, there are several constructive versions of the Intermediate Value Theorem, each of which can be successfully applied to most of the functions that arise in practice in analysis; see [3, pp. 40-41, 63; 11, pp. 54-58]. Moreover, the range and depth of the mathematics developed by Bishop in [2] is surely sufficient to disprove Hilbert's contention that.

Taking the principle of excluded middle from the mathematician would be the same, say, as proscribing the telescope to the astronomer or to the boxer the use of his fists. [20]

To end this introductory section, let us look at the role of various axioms of choice in constructive mathematics. We first observe that the full axiom of choice entails LEM [16], a fact that appears to conflict with Bishop's remark [2, p. 9] that the axiom of choice.. .is not a real source of nonconstructivity in classical mathematics. A choice function exists in constructive mathematics, because a choice is implied by the very meaning of existence.

Now, it is true that if to each element x of a set A their corresponds an element y of set B such that the property P(x, y) holds, then it is implied by the meaning of "existence" in constructive mathematics that there is a finite routine for computing an appropriate y ∈ B from a given x ∈ A; but this computation may depend not only on the value a but also on the information that shows that a belongs to the set A. The computation of the value at a of a function f from A to B would depend only on a, and not on the proof that a belongs to A; in other words, a function is extensional. So Bishop's remark is correct if he admits "functions" whose value depends on both a and a proof that a ∈ A, but is not correct if, as most constructive practitioners do, one only admits extensional functions.

Note that the axiom of choice will hold constructively if the set A is one for which no computation is necessary to demonstrate that an element belongs to it; Bishop calls such sets basic sets. Most constructivists would regard the set N^+ of positive integers as a basic set, a belief that is reflected in the acceptance of the principle of countable choice:

If to each positive integer n there corresponds an element y of a set A such that P(n, y), then there is a function $f : N^+ \rightarrow A$ such that P(n, f (n)) for each n ∈ N^+.

In fact, many constructive proofs use the stronger principle of dependent choice:

If a ∈ A, and to each x ∈ A their corresponds y ∈ A such that P(x, y), then there exists a function $f : N^+ \rightarrow A$ such that f (1) = a and P (f (n), f (n + 1)) for each n ∈ N^+.

CONSTRUCTIVE AXIOMS FOR THE REAL LINE

We now present a set of axioms for the constructive theory of the real line R, analogous to those given for the classical theory in [13]. Our axioms are intended to capture the idea that a real number is something that can be arbitrarily closely approximated by rational numbers. (In Bishop's formal construction, referred to above, that approximation is done by means of regular Cauchy sequences of rational numbers.)

It may, at first sight, seem strange to seek a constructive axiomatic presentation of the real line, but there are at least two reasons for wanting to do so. First, there is the possible advantage in teaching a course in constructive mathematics. Experience has convinced us that Bishop's formal constructive development of the real number system, starting with the notion of a regular (Cauchy) sequence of rational numbers, takes more time and effort than one might wish, and is sufficiently tricky for many students as to distract them from the main business of constructive analysis. We believe that it is possible to give a convincing motivation for the axioms presented below, by heuristic arguments based on the informal idea that a real number is something for which we can find (with emphasis on the word "find") arbitrarily close rational approximations.

Our second reason for seeking constructive axioms for the real line is the mathematician's natural curiosity about which properties suffice either to characterize that object or at least to enable one to develop real analysis constructively.

Our starting point is to assume the existence of a set R with

- A binary relation > (greater than)
- L a corresponding inequality relation # defined by

$$x \neq y \text{ if and only if } (x > y \text{ or } y > x)$$

- Binary operations $(x, y) \mapsto x + y$ (addition) and $(x, y) \mapsto xy$ (multiplication)
- Distinguished elements 0 (zero) and 1 (one) with $0 \neq 1$
- A unary operation $x \mapsto -x$
- A unary operation $x \mapsto x^{-1}$ on the set of elements $x \neq 0$.

The elements of R are called real numbers. We say that a real number x is positive if x >O, and negative if -x > 0. We define the relation ≥ (greuter than or equal to) by

$$x \neq y \text{ if and only if } \forall z \left(y > z \Rightarrow x > z \right)$$

and we define the relations < and ≤ in the usual way, calling x non-negative if x 2 0. Two real numbers x, y are equal if x ≥ y and y ≥ x, in which case we write x = y. Note that this notion of equality satisfies the usual properties of an equivalence relation.

We identify the sets N of natural numbers, N⁺ of positive integers, Z of integers, and Q of rational numbers with the usual subsets of R: for example, we identify N⁺ with {nl: n∈N⁺}.

$x + y = y + x,$

$\left(x + y \right) + z = x + \left(y + z \right),$

$0 + x = x,$

$x + \left(-x \right) = 0,$

$\left(xy \right) z = x \left(yz \right),$

$1x = x,$

$xx^{-1} = 1 \text{ if } x \neq 0,$

$x \left(y + z \right) = xy + xz.$

We also denote x^{-1} by $\dfrac{1}{x}$ or 1/x.

The answer is provided by a well-known example which shows that the statement

$$\forall x \in R \left(\neg (x = 0) \Rightarrow \exists y \in R \left(xy = 1 \right) \right)$$

is equivalent to Markov's Principle (MP):

$$\forall a \in (0,1)^{\mathbb{N}} \left(\neg (a = 0) \Rightarrow a \neq 0 \right)$$

that is, if (a_n) is a binary sequence such that $\neg \forall n (a_n = 0)$, then there exists n such that $a_n = 1$ (see [lo, Ch. 1, Problem 81). Since Markov's Principle is a form of unbounded search, it is not accepted by the majority of constructive mathematicians (although it is clearly true in classical mathematics).

The next group of axioms describes several properties of the ordering $>$.

R2. Basic properties of $>$:

$$\neg (x > y \text{ and } y > x).$$
$$\neg (x > y) \Rightarrow \forall z (x > z \vee z > y).$$
$$\neg (x \neq y) \Rightarrow x = y.$$
$$(x > y) \Rightarrow \forall z (x + z > y + z).$$
$$(x > 0 \wedge y > 0) \Rightarrow xy > 0.$$

The second of these axioms is a substitute for the law of trichotomy, and can be justified heuristically (see [9]). In connection with axiom R2(3), note that the statement

$$\forall x, y \in R \left(\neg (x = y) \Rightarrow x \neq y \right)$$

is equivalent to Markov's Principle [lo, Ch. 1, Problem 81.

Our last two axioms describe special properties of $>$ and $>$, For the second of these we need to know that the notions bounded above, bounded below, and bounded are defined as in classical mathematics; and that, for example, if S is a nonempty subset of R that is bounded above, then its least upper bound, if it exists, is the unique real number b such that

- b is an upper bound of S,
- for each b' <b there exists s∈S such that s > b'.

(Note that nonempty means what the intuitionists call "inhabited"; that is, we can construct an element of the set in question.)

R3. Special properties of >:

- Axiom of Archimedes: For each x∈R there exists n∈Z such that x <n.
- The least-upper-bound principle: Let S be a nonempty subset of R that is bounded above relative to the relation ≥, such that for all real numbers α, β ,with $\alpha > \beta$, either β is an upper bound of S or else there exists s∈S with s > X; then S has a least upper bound.

The first of these two axioms would seem to require no justification. The second is a little harder to motivate, but can be justified heuristically using an approximate interval halving argument [9].

Omitting many of the simpler proofs, we now derive some elementary constructive properties of the real line that involve the orderings > and ≥.

1. $\neg(x > x)$

2. $x \geq x$

3. $(x > y \wedge y > z \Rightarrow x > y)$

Since x > y, either x > z or z > y. The latter is ruled out by axiom R2 (1).

4. $(x > y \wedge y \geq x)$

If X> y and y≥x, then y>y, by definition of ≥. This is absurd.

5. $(x > y \geq z) \Rightarrow x > z$

Either x>z or z > y≥z. The latter is ruled out by 4.

6. $(x \geq y > z) \Rightarrow x > z$

   Use the definition of ≥.

7. $\neg(x > y) \Leftrightarrow y \geq z$

If $\neg(x > y)$ and $x > z$, then, by R2 (2), $y > z$. Hence $y \geq x$.

For the converse, use 4 above.

8. $\neg(x \geq y) \Leftrightarrow \neg\neg(y > x)$

(Note that the proposition $\neg(x \geq y) \Rightarrow y > x$ entails Markov's Principle.)

9. $(x \geq y \geq z) \Rightarrow x \geq z$

10. $(x \geq y \wedge y \geq x) \Rightarrow x = y$

If $x > y$, then $x > y \geq x$, which contradicts 4. Hence $\neg(x > y)$. Similarly, $\neg(y > x)$, so $\neg(x \neq y)$ and therefore, by axiom R2 (3), $x = y$.

11. $\neg(x > y \wedge x = y)$

12. $x \geq 0 \Rightarrow (x = 0 \Leftrightarrow \forall \varepsilon > 0(x < \varepsilon))$

If $x = 0$ and $\varepsilon > 0$, then either $\varepsilon > x$ or $x > 0$, and the latter is ruled out by the last result.

Conversely, suppose that $x < \varepsilon$ for all $\varepsilon > 0$. If $x > 0$, then $x < x$, a contradiction; so $0 \geq x$. Thus $x \geq 0$ and $0 \geq x$, and therefore $x = 0$.

13. $x + y > 0 \Rightarrow (x > 0 \vee y > 0)$

If $x + y > 0$, then either $x + y > x$ or else $x > 0$. In the former case, $y = (x+y) - x > x - x = 0$, by R2 (4).

14. $x > 0 \Rightarrow -x < 0$

$-x = 0 + (-x) < x + (-x) = 0$

15. $(x > y \wedge z < 0) \Rightarrow yz > xz$

Since -z>0, we have

-xz=x(-z)>y(-z)= -yz,

so

-xz+yz+xz>-yz+yz+xz

and therefore yz >xz.

16. $x \neq 0 \Rightarrow x^2 > 0$

17. $1 > 0$

For, $1^2 = 1 \neq 0$.

18. $x^2 > 0$

Suppose $x^2 < 0$. Then $\neg (x \neq 0)$, by 16; so $x = 0$ and therefore $x^2 = 0$, a contradiction.

Hence $\neg (x^2 < 0)$ and therefore $x^2 \geq 0$

19. $0 < x < 1 \Rightarrow x > x^2$

20. $-1 < x < 1 \Rightarrow \neg (x^2 > x \wedge x^2 > -x)$

Suppose that $-1 < x < 1$ and that $(x^2 > x \wedge x^2 > -x)$. If $x > 0$, then

$$-x = x(-1) < x^2 < x(1) = 0,$$

21. $x^2 > 0 \Rightarrow x \neq 0$

Either $x > 0$ or $x^2 > x$. In the latter case, either $x^2 > -x$ or $-x > x$. Suppose $--x > x$. Then

x-x>x+x=2x,

so $0 > x$. Thus we may assume that $x^2 > x$ and $x^2 > -x$. Then either $x > 0$ or $1 > x$. In the latter case, if $x > -1$, then we contradict 20; so $0 > -1 \geq x$ and therefore $0 > x$.

22. $x > 0 \Rightarrow x^{-1} \geq 0$

If $0 > x^{-1}$, then $1 = x^{-1}x < x^{-1}0 = 0$, a contradiction.

23. Let m, m' be integers, and n, n' positive integers. Then $m/n > m'/n'$ if and only if $mn' > m'n$.

    This is an exercise in algebra, starting from the fact that $1 > 0$

24. If $n \in N^+$, then $n^{-1} > 0$.

25. If $x > 0$ and $y \geq 0$, then there exists $n \in Z$ such that $nx > y$. Using axiom R3 (1), choose $n \in Z$ such that $n > n^{-1} y$. Then $nx > (x^{-1} y) x = y \geq 0$, so $n > 0$.

26. $x > 0 \Rightarrow x^{-1} > 0$

    Using axiom R3 (1), choose $n \in Z$ such that $n > x$. Then $n > 0$ and so, by axiom R2 (5), $nx > 0$. Suppose that $x^{-1} < (nx) n^{-1} = x$,

    a contradiction. Hence $x^{-1} \geq n^{-1} > 0$, and therefore $x^{-1} > 0$

27. $xy > 0 \Rightarrow (x \neq 0 \wedge y \neq 0)$

    Since

    $$(x + y)^2 + 2xy + y^2 \geq 2xy > 0,$$

    we see that $x + y \neq 0$. Without loss of generality, take $x + y > 0$. Then either $x > 0$ or $y > 0$. Taking, for example, the first case, since $x^{-1} > 0$ we have

    $y = x^{-1}(xy) > x^{-1} 0 = 0$.

28. If $x > 0$, then there exists $n \in N^+$ such that $x < n < x + 2$. Choose $m \in N^{+1}$ such that Either $x + 1 < m$ or $m < m - 1$. Repeatedly using this procedure, we either reach the desired conclusion or we show that $x < 1$. In the latter case, either $x + 1 < 1$ and therefore $x < 0$, which is absurd; or, as must be the case, $1 < x + 2$.

29. If $a < b$, then there exists $r \in Q$ such that $a < r < b$. First assume that $a > 0$. Using 25, choose $n > 0$ such that $n(b - a) > 2$ and therefore $na < na + 2 < nb$. By the preceding result, there exists $m \in N^{+1}$ such that $na < m < na + 2 < nb$. Then $a < m/n < b$, so we can take $r = m/n$. In the general case, choose $n \in Z$ such that $-a < n$ Then $b + n > a + n > 0$; so, by the first part of the proof, there exists a rational number r' with $a + n < r' < b + n$. Then $r = r' - n$ is rational and $a < r < b$.

There is an obvious question relating to our system of axioms for R: is it categorical? In other words, are any two models of these axioms isomorphic? The answer is perhaps at first sight surprising: there are nonisomorphic models of the axioms. For example, the Dedekind reals satisfy all our axioms and a stronger form of the least-upper-bound principle, and so provide a model that is not isomorphic to the one obtained by Bishop's construction of R using regular sequences of rational numbers (see [26, pp. 270-2771).

THE COMPLETENESS OF R

We now deal with some consequences of the least-upper-bound principle that will lead us to a proof of the Cauchy-sequence completeness of R.

A set S is said to be

- Finitely enumerable if there exist a natural number n and a mapping of { 1,. . . , n} onto S;
- Finite if there exist a natural number n and a oneeone mapping of { 1,. . . , n} onto S. In the first case we also say that S has at most n elements, and in the second that S has (exactly) n elements.

The statement

Every finitely enumerable subset of R is finite

entails LPO: given a binary sequence (a,), consider the finitely enumerable set

$$\left\{0, \sum_{n=1}^{\infty} 2^{-n} a_n\right\}.$$

Lemma 1: If S is a finitely enumerable subset of R, then sups and inf S exist.

Proof: Let $S = \{x_1, \ldots, x_n\}$. Given real numbers α, β with $\alpha > \beta$ we apply axiom R2 (2) n times to prove that either $x_k < \beta$ for each k or else there exists j such that $xj > \alpha$. Axiom R3 (2) now shows that sup S exists. The proof for inf S is similar.

Let S be a subset of R. By an E-approximation to S we mean a subset F of S such that for each $x \in S$ there exists $y \in F$ with $|x - y| < \varepsilon$. We say that S is totally bounded if for each $\varepsilon > 0$ there exists a finite s-approximation to S. It is an exercise to show that a set is totally bounded if (and clearly only if) for each $\varepsilon > 0$ it contains a finitely enumerable ε-approximation.

Proposition 1: If S is a totally bounded subset of R, then sup S and inf S exist.

Proof: Let α, β be real numbers with $\alpha > \beta$ and write $\varepsilon = \frac{1}{2}(\beta - \alpha)$. There exists a finite ε-approximation $\{x_1, \ldots, x_n\}$ to S. By the preceding lemma,

$\sigma = \sup \{x_1, \ldots, x_n\}$

exists. Either $\sigma > \alpha$ or $\sigma < \alpha + \varepsilon$ In the first case there exists j such that $x_j > \alpha$. In the second, consider any $x \in S$. Choosing j such that $|x - x_j| < \varepsilon$, we have

$$x \le x_j + |x - x_j| < \sigma + \varepsilon < \alpha + 2\varepsilon = \beta.$$

So in this case, β is an upper bound for S. It follows from axiom R3 (2) that sups exists. Similarly, inf S exists.

A sequence (x_n) of real numbers is said to be

- A Cauchy sequence if for each $\varepsilon > 0$ there exists N such that $|x_m - x_n| < \varepsilon$ all $m, n \ge N$;
- Convergent if there exists a real number l, called the limit of (x_n), such that for each $\varepsilon > 0$ there exists N such that $|x_n - l| < \varepsilon$ for all $n \ge N$.

We aim to prove the converse of the easily established result that a convergent sequence in R is a Cauchy sequence

Lemma 2: if (x_n) is a Cauchy sequence in R, then $\{x_n : n \ge 1)$ is totally bounded.

Proof: Given $\varepsilon > 0$, choose N such that $|x_m - x_n| < \varepsilon$ for all $m, n \geq N$. Then $\{x_1, \ldots, x_N\}$ is an s-approximation to $\{x_n : n \geq 1)$.

Lemma 3: If a Cauchy sequence (x_n) in R has a convergent subsequence $\left(x_{n_k}\right)$, then $\lim_n \to \infty x_n$ exists and equals $\lim_k \to \infty x_{n_k}$.

Proof: Let $l = \lim_k \to \infty x_{n_k}$. Given $\varepsilon < 0$ choose K such that $\left|l - x_{n_k}\right| < \varepsilon$ for all $k \geq K$ Now choose $N > K$ such that $|x_m - x_n| < \varepsilon$ for all $m, n \geq N$. Then as $n_N \geq N$, we have

$$\left|l - x_k\right| \leq \left|l - x_{n_N}\right| + \left|x_{n_N} - x_k\right| < \varepsilon + \varepsilon = 2\varepsilon.$$

Theorem. R is complete: that is, every Cauchy sequence in R converges.

Proof. Let (x_n) be a Cauchy sequence in R. Then for each n, (x_k) $k \geq n$ is a Cauchy sequence, so by Lemma 2 and Proposition 1,

$$S_n = \sup \{x_k : k \geq n\}$$

exists. Given $\varepsilon < 0$ choose N_ε such that $|x_m - x_n| < \varepsilon$ for all $m, n \geq N_\varepsilon$. For such m and n, suppose that

$$S_j > S_m - \left(S_m - S_n - \varepsilon\right);$$

then

$$x_j - x_n > S_n + \varepsilon - S_n = \varepsilon,$$

which is absurd as $|x_j - x_n| < \varepsilon$. Hence $|S_m - S_n| \leq \varepsilon$ for all $m, n \geq N_\varepsilon$, and so (s_n) is a Cauchy sequence. Thus, by Lemma 2 and Proposition 1,

$$t = \inf_{n \geq 1} S_n$$

exists.

We show that $\lim_{n \to \infty} x_n = t$. To this end, choose N such that $S_N < t + \varepsilon$. Then as

$$S_1 \geq S_2 \cdots \geq \inf_{n \geq 1} S_n,$$

for each $n \geq N$ we have $t - \varepsilon < S_n < t + \varepsilon$, so there exists $v \geq n$ with $t - \varepsilon < x_v < t + \varepsilon$. A simple inductive construction now provides a subsequence $\left(x_{n_k}\right)_{k=1}^{\infty}$ of (x_n) such that $t - k^{-1} < x_{n_k} < t + k^{-1}$ for each k. Then $\lim_{k \to \infty} x_{n_k} = t$. It follows from Lemma 3 that (x_n) converges to t.

REFERENCES

1. . M. Beeson, Foundations of Constructive Mathematics, Springer, Heidelberg, 1985.
2. E. Bishop, Foundations of Constructive Analysis, McGraw-Hill, New York, 1967.
3. E. Bishop, D. Bridges, Constructive Analysis, Grundlehren Math. Wiss., vol. 279, Springer, Heidelberg, 1985.
4. D. Bridges, Constructive mathematics, its set theory and practice, D.Phil. Thesis, University of Oxford, 1975.
5. D. Bridges, Constructive mathematics and unbounded operators - a reply to Hellman, J. Philos. Logic 25 (1995) 549-561.
6. D. Bridges, Constructive Truth in Practice, in: H.G. Dales, G. Oliveri (Eds.), Truth in Mathematics, Proc. Conf. held at Mussomeli, Sicily, 13-21 September 1995, Oxford University Press, Oxford, 1998, to appear.
7. D. Bridges, A constructive Morse theory of sets, in: D. Skordev (Ed.), Mathematical Logic and its Applications, Plenum Publishing Corp., New York, 1987, pp. 61-79.
8. D. Bridges, 0. Demuth, On the Lebesgue measurability of continuous functions in constructive analysis, Bull. Amer. Math. Sot. 24(2) (1991) 259-276.
9. D. Bridges, S. Reeves, Constructive mathematics in theory and programming practice, Philos. Math. (1999), to appear.
10. D. Bridges, F. Richman, Varieties of Constructive Mathematics, London Math. Sot. Lecture Notes, vol. 97, Cambridge University Press, Cambridge, 1987.
11. L.E.J. Brouwer, Over de Grondslagen der Wiskunde, Doctoral Thesis, University of Amsterdam, 1907. Reprinted with additional material (D. van Dalen, Ed.) by Matematisch Centrum, Amsterdam, 1981.
12. R.L. Constable et al., Implementing Mathematics with the Nuprl Proof Development System, PrenticeHall, Englewood Cliffs, NJ, 1986.

13. J. Dieudonnt, Foundations of Modem Analysis, Academic Press, New York, 1960.
14. S. Feferman, Constructive theories of functions and classes, in: M. Boffa, D. van Dalen, K. McAloon (Eds.), Logic Colloquium '78: Proceedings of the Logic Colloquium at Mom., 1978, North-Holland, Amsterdam, 1979, pp. 159-224. .
15. H. Friedman, Set theoretic foundations for constructive analysis, Ann. Math. 105 (1977) l-28.
16. N.D. Goodman, J. Myhill, Choice implies excluded middle, Zeit. Logik Grundlagen Math. 24 (1978) 461.
17. S. Hayashi, H. Nakano, PX: A Computational Logic, MIT Press, Cambridge, MA, 1988.
18. G. Hellman, Constructive mathematics and quantum mechanics: unbounded operators and the spectral theorem, J. Philos. Logic 22 (1993) 221-248.
    A Heyting, Die formalen Regeln der intuitionistischen Logik, Sitzungsber. Preuss. Akad. Wiss. Berlin (1930) 42-56.
19. D. Hilbert, Die Grundlagen der Mathematik, Hamburger Mathematische Einzelschritben 5, Teubner, Leipzig, 1928. Reprinted in English translation in .27
20. , in which the exact quotation appears on p. 476.
21. P. Martin-Liif, An intuitionistic theory of types: predicative part, in: H.E. Rose, J.C. Shepherdson (Eds.), Logic Colloquium 1973, North-Holland, Amsterdam, 1975, pp. 73-l 18.
22. J. Myhill, Some properties of intuitionistic Zennelo-Fraenkel set theory, in: A. Mathias, H. Rogers (Eds.), Cambridge Summer School in Mathematical Logic, Lecture Notes in Mathematics, vol. 337, Springer, Berlin, 1973, pp. 206-231.
23. J. Myhill, Constructive set theory, J. Symbolic Logic 40(3) (1975) 347-382.
24. F. Richman, Intuitionism as generalization, Philos. Math. 5 (1990) 124-128 (MR #9lg:O3014).
25. F. Richman, Interview with a constructive mathematician, Modem Logic 6 (1996) 247-271.
26. A.S. Troelstra, D. van Dalen, Constructivity in Mathematics: An Introduction (two volumes), NorthHolland, Amsterdam, 1988.
27. J. van Heijenoort, From Frege to Gijdel, A Source Book in Mathematical Logic 1879-193 1, Harvard University Press, Cambridge, MA, 1967.

CITATION

Douglas S. Bridges, Constructive mathematics: a foundation for computable analysis, Theoretical Computer Science, Volume 219, Issues 1–2, 28 May 1999, Pages 95-109, ISSN 0304-3975, http://dx.doi.org/10.1016/S0304-3975(98)00285-0.

Numerical Approximation of Real Finite Nonnegative Function by the Modulus of Discrete Fourier Transform

Petro Savenko and Myroslava Tkach

Pidstryhach Institute for Applied Problems of Mechanics and Mathematics, National Academy of Sciences of Ukraine, Lviv, Ukraine

9

ABSTRACT

The numer ical algorithms for finding the lines of branching and branching-off solutions of nonlinear problem on mean-square approximation of a real finite nonnegative function with respect to two variables by the modulus of double discrete Fourier transform dependent on two parameters, are constructed and justified.

INTRODUCTION

The mean-square approximation of real finite nonnegative function with respect to two variables by the modulus of double discrete Fourier transform dependent on physical parameters, is widely used, in particular, at modeling and solution of the synthesis problems of different types of antenna arrays, signal processing etc. [1-3]. Nonuniqueness and branching of solutions are essential features of nonlinear approximation problem which remains unexplored. The problem on finding the set of branching points, in turn, is not adequately explored nonlinear spectral two-parametric problem. The methods of investigation and numerical finding the solutions of one-parametric spectral problems at presence of discrete spectrum [4-8] are most well-developed. The existence of coherent components of spectrum, which are spectral lines for the case of real parameters [9], is essential difference of nonlinear two-parametric spectral problems.

In the work a variational problem on the best meansquare approximation of a real finite nonnegative function by the module of double discrete Fourier transform is reduced to finding the solutions of Hammerstein type nonlinear two-dimensional integral equation. Using the Schauder principle the existence of solutions is proved. The existence theorem of coherent components of spectrum of holomorphic matrix functions dependent on two spectral parameters is proved. It justifies the application of implicit functions methods to multiparametric spectral problems [9]. The applicability of this theorem to the analysis of spectrum of two-dimensional integral homogeneous equation to which is reduced the problem on finding the lines of possible branching of solutions of the Hammerstein equation, is shown. Algorithms for numerical finding the optimum solutions of an approximation problem are constructed and justified. Numerical examples are presented

PROBLEM FORMULATION, BASIC EQUATIONS AND RELATIONS

Consider the special case of double discrete Fourier transform

$$f(s_1, s_2) = \sum_{n=-N_1}^{N_2} \sum_{m=-M_1(n)}^{M_2(n)} I_{nm} \exp\left[i\left(\tilde{c}_1 x_{nm} s_1 + \tilde{c}_2 y_{nm} s_2\right)\right]$$

setting here

$x_{nm} = n\Delta_x \ (n = -N \div N), y_{nm} = m\Delta_y \ (m = -M \div M); c_1 = \tilde{c}_1 \Delta_x, c_2 = \tilde{c}_2 \Delta_y.$
If it is necessary for the accepted assumptions we shall consider the formula

$$f(s_1, s_2) = A\mathbf{I} \equiv \sum_{n=-N}^{N} \sum_{m=-M}^{M} I_{nm} \exp\left[i\left(c_1 n s_1 + c_2 m s_2\right)\right]$$

(1)

as a linear operator, acting from complex finite-dimensional space $H_1 = \mathbb{C}^{N_2 \times M_2}$ ($N_2 = 2N+1, M_2 = 2M+1$) into the space of complex-valued continuous functions with respect to two real variables determined in the domain

$$\Omega = \left\{ (s_1, s_2) : |s_1| \leq \pi/c_1, \ |s_2| \leq \pi/c_2 \right\}.$$

Here c_1, c_2 are any real non-dimensional numerical parameters belonging to

$$\Lambda_c = \left\{ (c_1, c_2) : 0 < c_1 \leq a, \ 0 < c_2 \leq b \right\}.$$

The function $f(s_1, s_2)$ is $2\pi/c_1$- periodic function on argument s_1 and $2\pi/c_2$- periodic on s_2.

In considered spaces we introduce scalar products and generable by them norms

$$\left(\mathbf{I}_1, \mathbf{I}_2 \right)_{H_1} = \frac{4\pi^2}{c_1 c_2} \sum_{n=-N}^{N} \sum_{m=-M}^{M} I_{nm} \overline{I_{nm}},$$

$$\|\mathbf{I}\| = \left(\mathbf{I}, \mathbf{I} \right)_{H_1}^{1/2},$$

(2)

$$\left(f_1, f_2 \right)_{C_{(\Omega)}^{(2)}} = \iint_{\Omega} f_1(s_1, s_2) \overline{f_2(s_1, s_2)} ds_1 ds_2,$$

$$\|f\| = \left(f, f \right)_{C_{(\Omega)}^{(2)}}^{1/2}.$$

(3)

Denote an augmented space of continuous functions with entered scalar product and norm (3) as $C_{(\Omega)}^{(2)}$ and notice that its augmentation coincides with the Hilbert space $L_2(\Omega)$ [10].

By direct check we are sure that such equality

$$\left\|A\mathbf{I}\right\|^2 = \iint_{\Omega}\left|f\left(s_1,s_2\right)\right|^2 ds_1 ds_2 = \sum_{n,m}\left|I_{nm}\right|^2 = \left\|\mathbf{I}\right\|^2$$

(4)

is valid. From here follows, that A is isometric operator in sense (4).

Using the entered scalar products (2) and (3) we find the conjugate operator required later on

$$A^* f = \frac{c_1 c_2}{4\pi^2}\iint_{\Omega} f\left(s_1,s_2\right)\exp\left[-i\left(c_1 n s_1 + c_2 m s_2\right)\right] ds_1 ds_2$$

$$\left(n = -N \div N,\ m = -M \div M\right).$$

(5)

Let such function be given

$$\tilde{F}(s_1,s_2) = \begin{cases} F(s_1,s_2), & (s_1,s_2) \in \overline{G} \subseteq \Omega, \\ 0, & (s_1,s_2) \in \Omega \setminus \overline{G}, \end{cases}$$

(6)

where $F(s_1, s_2)$ is a real continuous and nonnegative in the domain $\overline{G}$ function.

Consider a problem on the best mean-square approximation of the function 1 2 F(s s,) in the domain W by the module of double discrete Fourier transform (1) owing to select of coefficients of the vector I . We shall formulate it as a minimization problem of the functional

$$\sigma(\mathbf{I}) = \left\|F - \left|A\mathbf{I}\right|\right\|^2_{L^{(2)}_{(\Omega)}} = \left\|F - \left|f\right|\right\|^2_{L^{(2)}_{(\Omega)}}$$

(7)

in the Hilbertian space H_1. Taking into account (4) and (5), we write the functional $\sigma(I)$ in a simplified form

$$\sigma(\mathbf{I}) = \|F\|^2_{C^{(2)}_{(\Omega)}} - 2\left(F,|A\mathbf{I}|\right)^2_{C^{(2)}_{(\Omega)}} + \|\mathbf{I}\|^2_{H_1}.$$

(8)

On the basis of necessary condition of functional minimum we obtain a nonlinear system of equations relating to the components of vector I in the space H_1 that are represented in the vector and expanded forms, respectively:

$$\mathbf{I} = A^*\left\{F \exp\left[i\arg\left(A\mathbf{I}\right)\right]\right\},$$

(9)

$$I_{nm} = \frac{c_1 c_2}{4\pi^2} \iint_{\Omega} F\left(s_1, s_2\right) \exp\left\{i\left[\arg\left(\sum_{k=-N}^{N}\sum_{l=-M}^{M} I_{nm} \times\right.\right.\right.$$
$$\times \exp\left[i\left(c_1 k s_1 + c_2 l s_2\right)\right]\right) - \left(c_1 n s_1 + c_2 m s_2\right)\left.\right]\right\} ds_1 ds_2$$
$$\left(n = -N \div N,\ m = -M \div M\right).$$

(9)

Acting on both parts of (9) by operator A we obtain equivalent to (9) the Hammerstein type nonlinear integral equation relating to f:

$$f(Q) = \mathbf{B}f \equiv \iint_{\Omega} K\left(Q,Q',\mathbf{c}\right) F(Q') \exp\left[i\arg f(Q')\right] dQ',$$

(10)

where $Q' = \left(s_1, s_2\right), dQ' = ds_1 ds_2, \mathbf{c} = \left(c_1, c_2\right);$

$$K\left(Q,Q',\mathbf{c}\right) = K_1\left(s_1, s_1', c_1\right) \cdot K_2\left(s_2, s_2', c_2\right),$$

(11)

$$K_1\left(s_1, s_1', c_1\right) = \frac{c_1}{2\pi} \sum_{n=-N}^{N} \exp\left[ic_1 n\left(s_1 - s_1'\right)\right] \equiv$$

$$\equiv \frac{c_1}{2\pi} \frac{\sin\left[\dfrac{N_2 c_1}{2}\left(s_1 - s_1'\right)\right]}{\dfrac{c_1}{2}\left(s_1 - s_1'\right)},$$

$$K_2\left(s_2, s_2', c_2\right) = \frac{c_2}{2\pi} \sum_{m=-M}^{M} \exp\left[ic_2 m\left(s_2 - s_2'\right)\right] \equiv$$

$$\equiv \frac{c_2}{2\pi} \frac{\sin\left[\dfrac{M_2 c_2}{2}\left(s_2 - s_2'\right)\right]}{\dfrac{c_2}{2}\left(s_2 - s_2'\right)}.$$

Note, that the kernel (11) of Equation (10) is degenerate and real.

We shall consider one of the properties of function exp(i arg f(Q')) entering into (10) at f $f(Q') \to 0$. Obviously that the function

$$\exp\left(i \arg f(Q')\right) = \frac{f(Q')}{|f(Q')|} = \frac{u(Q') + iv(Q')}{\left(u^2(Q') + v^2(Q')\right)^{1/2}}$$

is continuous if $u(Q') = \mathrm{Re} f(Q')$ and $v(Q') = \mathrm{Im} f(Q')$ are continuous functions, where $\left|\exp\left(i \arg f(Q')\right)\right| = 1$ for any $f(Q')$. If $u(Q') \to 0$ and $v(Q') \to 0$ simultaneously then $f(Q') \to 0$ is a complex zero. Its argument is undetermined accordingly to definition of complex zero [11, p. 20]. On this basis we redefine $\exp_{(i \arg f(Q'))}$ at $u(Q') \to 0$ and $v(Q') \to 0$ as a function which has module equal to unit and undetermined argument.

The equivalence of (9) and (10) follows from the following lemma.

Lemma 1: Between solutions of Equations (9) and (10) there exists bijection, i.e., if I_* is a solution of (9) then $f_* = AI_* = $ is the solution of (10); on the contrary, if f_* is the solution of (10) then

$$I_* = A^* \left\{ F \exp\left[i \arg(f_*) \right] \right\}$$

(12)

is the solution of (9).

Proof: Let I_* be a solution of (9). Then $I_* - A^* \left\{ F \exp\left[i \arg(AI_*) \right] \right\} \equiv 0$. Acting on this identity by the linear operator A, we have $AI_* - AA^* \left\{ F \exp\left[i \arg(AI_*) \right] \right\} \equiv 0$. Since the operator A acts from the space $H_1 = \mathbb{C}^{N_2 \times M_2}$ into the space $C_{(\Omega)}^{(2)}$ and accordingly into the space $\tilde{H}_f = L_2(\Omega)$, and the set of its nulls consists of only null element from the last identity follows, what $AI_* = f_* \in \tilde{H}_f$ is a solution of (10).

On the contrary, let $f_* \in \tilde{H}_f$ solves the Equation (10). The operator A^* acts from the space $\tilde{H}_f^* = L_2^*(\Omega)$ into the space $H_1 = \mathbb{C}^{N_2 \times M_2}$ [10] and the Hilbertian space L_2^* coincides with the space L_2 [10]. From here follows, that A^* acts from the space $\tilde{H}_f = L_2(\Omega)$ into the space $H_1 = \mathbb{C}^{N_2 \times M_2}$. Taking into account that F is a finite function determined by (6), and f_* is continuous, the function $F \exp(i \arg(f_*))$ is quadratic integrability in the domain Ω, i.e. $F \exp(i \arg(f_*)) \in H_f$. Thus $A^* \left(F \exp(i \arg(f_*)) \right) = I_* \in H_1$ and the right part of (10) is the result of action of operator A on an element I_*, i.e. $AI_* = AA^* \left(F \exp(i \arg(f_*)) \right) = f_*$. Writing this equality as $A \left(I_* - A^* \left(F \exp(i \arg(AI_*)) \right) \right) = 0$ and taking into account that a set of operator nulls consists of only a null element we obtain $I_* = A^* \left(F \exp(i \arg(AI_*)) \right)$. So, $I_* = A^* \left(F \exp(i \arg(f_*)) \right)$ solves the Equation (9). Lemma is proved.

Thus owing to the equivalence of (9) and (10) we consider simpler of them, namely (10). The Equation (9) is a more complicated equation in sense that in its right part the operator A is in an index of the power of exponent.

Besides taking into account that a set of values of operator A is a set of continuous functions in the domain (Ω) belonging to the space L_2 (Ω) and this set is a compact in the space L_2 (Ω) [12], we shall investigate solutions of (10) in the space $C(\Omega)$.

Formulate the important properties of (10), which are checked directly.

1) If function $f(Q)$ is a solution of (10) then the conjugate complex function $\overline{f(Q)}$ is also the solution of (10).
2) If function $f(Q)$ is a solution of (10), then $\exp(i\beta)$ $f(Q)$ is also the solution of (10) (β is any real constant).
3) For even on two arguments (or on one argument) functions $F(s_1,s_2)$ the nonlinear operator B that is in the right part of (10), is an invariant concerning the type of parity of the function arg f (s_1,s_2) on two arguments (or on that argument on which $F(s_1,s_2)$ is an even function).

Below taking into account the property 2) for uniqueness of solutions we set the parameter $\beta = 0$.

Consider the operator

$$Df \equiv \iint_{\Omega} K(Q,Q',\mathbf{c}) f(Q')dQ'$$

(13)

and corresponding to it quadratic form

$$(Df,f) = \iint_{\Omega} \iint_{\Omega} K(Q,Q',\mathbf{c}) f(Q')dQ' \overline{f(Q)}dQ =$$

$$= \sum_{n=-N}^{N} \sum_{m=-M}^{M} \left| \iint_{\Omega} \exp\left[i\left(c_1 n s_1 + c_2 m s_2\right)\right] \overline{f(s_1,s_2)} ds_1 ds_2 \right|^2 =$$

$$= \left(\frac{2\pi}{c_1}\right)\left(\frac{2\pi}{c_2}\right)\|\mathbf{I}\|^2 \geq 0.$$

Obviously that this inequality modifies into equality only as $I = 0$. From here follows that the kernel $K(Q,Q',c)$ is positively defined [13]. Accordingly operator D is positive on nonnegative functions cone K of the space $C(\Omega)$ [14]. According to it D leaves invariant the cone K, i.e. $DK \subset$.

Complex decomplexified space $C(\Omega)$ W [10] we consider as a direct sum of two real spaces of continuous functions $C(\Omega) = C(\Omega) \otimes C(\Omega)$ in the domain Ω. The elements of this space are written as $f = (u, v)^{\mathsf{T}} \in C(\Omega), u \in C(\Omega), v \in C(\Omega)$. Norms in these spaces have the form:

$$\|u\|_{C(\Omega)} = \max_{Q \in \Omega} |u(Q)|, \quad \|v\|_{C(\Omega)} = \max_{Q \in \Omega} |v(Q)|,$$

$$\|f\|_{C(\Omega)} = \max\left(\|u\|_{C(\Omega)}, \|v\|_{C(\Omega)}\right).$$

The Equation (10) in the decomplexified space $C(\Omega)$ we reduce to equivalent to it system of the nonlinear equations

$$u(Q) = B_1(u,v) \equiv \iint_{\Omega} K(Q,Q',c) F(Q') \frac{u(Q')}{\sqrt{u^2(Q') + v^2(Q')}} dQ',$$

$$v(Q) = B_2(u,v) \equiv \iint_{\Omega} K(Q,Q',c) F(Q') \frac{v(Q')}{\sqrt{u^2(Q') + v^2(Q')}} dQ'. \tag{14}$$

Denote the closed convex set of continuous functions as $S_R \subset C(\Omega)$ supposing that

$$S_R = S_{R_u} \oplus S_{R_v}, \quad S_{R_u} = \left\{ u : \|u\|_{C(\Omega)} \leq R \right\},$$

$$S_{R_v} = \left\{ v : \|v\|_{C(\Omega)} \leq R \right\},$$

$$R = \max_{Q \in \Omega} \iint\limits_{\Omega} |K(Q,Q',\mathbf{c})| F(Q') dQ'.$$

Theorem 1: The operator $B = (B_1, B_2)^\mathsf{T}$ determined by the Formula (14) maps a closed convex set S_R of the Banach space $C(\Omega)$ in itself and it is completely continuous.

Proof: At first we show that $B : C(\Omega) \to C(\Omega)$. Let $f = (u,v)^\mathsf{T}$ be any function belonging to $C(\Omega)$. At $(c_1, c_2) \in \wedge_c$ the kernel $K(Q,Q',c)$ is a continuous function with respect to its arguments in the closed domain $\Omega \times \Omega$. Then accordingly to the Cantor theorem [15] $K(Q,Q',c)$ is a uniformly continuous function in $\Omega \times \Omega$. From here follows: for any points $(Q_1, Q'_1), (Q_2, Q'_2)$ such that whenever $|(Q_1,Q'_1) - (Q_2,Q'_2)| < \delta$, then $|K(Q_1,Q'_1) - K(Q_2,Q'_2)| < \dfrac{\varepsilon}{\alpha}$, where $\alpha = \iint\limits_{\Omega} F(Q') dQ'$. On this basis we obtain

$$|u(Q_1) - u(Q_2)| = \left| \iint\limits_{\Omega} F(Q') \left[K(Q_1,Q',\mathbf{c}) - K(Q_2,Q',\mathbf{c}) \right] \times \right.$$

$$\times \frac{u(Q')}{\sqrt{u^2(Q')+v^2(Q')}}\, dQ'\Bigg| \le \frac{\varepsilon}{a}\iint_G F(Q')dQ' = \varepsilon,$$

$$\text{since} \quad \max_{Q'\in\bar{\Omega}}\left|\frac{u(Q')}{\sqrt{u^2(Q')+v^2(Q')}}\right| \le 1.$$

Analogously we have that $\left|v(Q_1)-v(Q_2)\right|\le\varepsilon$ whenever $\left|(Q_1,Q_1')-(Q_2,Q,\text{ i.e. }(u,v)^T \in C(\Omega)\text{ and }B:C(\Omega)\to C(\Omega)\right|$.

To prove the property of a complete continuity of the operator $B=(B_1,B_2)^T$ it is necessary to prove its compactness and continuity [12]. Show a continuity $B=(B_1,B_2)^T$. Let $f_1=(u_1,v_1)^T \in S_R$ be any fixed function and $f_2=(u_2,v_2)^T$ be any function belonging to S_R. It is necessary to show that $\left\|Bf_1 - Bf_2\right\|_{C(\Omega)} \to 0$ as $\left\|f_1 - f_2\right\|_{C(\Omega)} \to 0$. Set $u_2 = u_1 + \Delta u, v_2 = v_1 + \Delta v$. Taking into account these equalities we obtain

$$\frac{u_2}{\sqrt{u_2^2+v_2^2}} = \frac{u_1+\Delta u}{\sqrt{u_1^2+v_1^2}\sqrt{1+\dfrac{2u_1\Delta u + 2v_1\Delta v + \Delta u^2 + \Delta v^2}{u_1^2+v_1^2}}}.$$

At $\left\|\Delta u\right\|_{C(\Omega)} \to 0, \left\|\Delta v\right\|_{C(\Omega)} \to 0$ we have

$$\lim_{\substack{\|\Delta u\|\to 0,\\ \|\Delta v\|\to 0}}\left\|\frac{u_1(Q)}{\sqrt{u_1^2(Q)+v_1^2(Q)}}-\frac{u_2(Q)}{\sqrt{u_2^2(Q)+v_2^2(Q)}}\right\|_{C(\Omega)} \le$$

$$\le \lim_{\substack{\|\Delta u\|\to 0, \, Q\in\Omega\\ \|\Delta v\|\to 0}}\max\left\{\left|\frac{u_1(Q)}{\sqrt{u_1^2(Q)+v_1^2(Q)}}\left(1-\frac{1}{P(u_1(Q),v_1(Q))}\right)\right|+\right.$$

$$+ \left| \frac{\Delta u(Q)}{\sqrt{u_1^2(Q) + v_1^2(Q)} P\big(u_1(Q), v_1(Q)\big)} \right| \Bigg\} = 0 \, ,$$

(16)

where

$$P\big(u_1(Q), v_1(Q)\big) =$$

$$= \sqrt{1 + \frac{2u_1(Q)\Delta u(Q) + 2v_1(Q)\Delta v(Q) + \Delta u^2(Q) + \Delta v^2(Q)}{u_1^2(Q) + v_1^2(Q)}} \, ,$$

since

$$\lim_{\substack{\|\Delta u\| \to 0, \\ \|\Delta v\| \to 0}} \max_{Q \in \Omega} \left| P\big(u_1(Q), v_1(Q)\big) \right| = 1 \, .$$

Similarly we obtain

$$\lim_{\substack{\|\Delta u\| \to 0, \\ \|\Delta v\| \to 0}} \max_{Q \in \Omega} \left| \frac{v_1(Q)}{\sqrt{u_1^2(Q) + v_1^2(Q)}} - \frac{v_2(Q)}{\sqrt{u_2^2(Q) + v_2^2(Q)}} \right| = 0 \, .$$

(17)

Thus, from (16) and (17) follows

$$\lim_{\substack{\|\Delta u\| \to 0, \\ \|\Delta v\| \to 0}} \left\| B_1(u_1, v_1) - B_1(u_2, v_2) \right\|_{C(\Omega)} =$$

$$= \lim_{\substack{\|\Delta u\| \to 0, \\ \|\Delta v\| \to 0}} \max_{Q \in \Omega} \left| \iint_{\Omega} F(Q') K(Q, Q', \mathbf{c}) \times \right.$$

$$\times \left. \left(\frac{u_1(Q')}{\sqrt{u_1^2(Q') + v_1^2(Q')}} - \frac{u_2(Q')}{\sqrt{u_2^2(Q') + v_2^2 dQ'(Q')}} \right) dQ' \right| = 0 \, .$$

Analogously

$$\lim_{\substack{\|\Delta u\|_{C(\Omega)} \to 0, \\ \|\Delta v\|_{C(\Omega)} \to 0}} \left\| B_2(u_1, v_1) - B_2(u_2, v_2) \right\|_{C(\Omega)} = 0.$$

So, $B = (B_1, B_2)^\mathsf{T}$ is continuous operator from $C(\Omega)$ into $C(\Omega)$.

We show that a set of functions $S_g = BS_R$ satisfies conditions of the Arzela theorem [12], i.e. we show that functions of the set S_g are uniformly bounded and equipotentially continuous. Furthermore $BS_R \subset S_R$. Let $g = (w, \omega)^\mathsf{T} = Bf \equiv (B_1(u,v), B_2(u,v))^\mathsf{T}$, where $f = (u, v)^\mathsf{T}$ is any function of the set S_R. Then as $|(Q_1, Q') - (Q_2, Q')| < \delta$ analogously with (15) we have

$$\begin{pmatrix} |w(Q_1) - w(Q_2)| \\ |\omega(Q_1) - \omega(Q_2)| \end{pmatrix} \le \begin{pmatrix} \dfrac{\varepsilon}{a} \iint_\Omega F(Q')dQ' \\ \dfrac{\varepsilon}{a} \iint_\Omega F(Q')dQ' \end{pmatrix} = \begin{pmatrix} \varepsilon \\ \varepsilon \end{pmatrix}.$$

Thus functions of the set $S_g = BS_R$ are equipotentially continuous. The uniform boundedness of the set g $S_g = BS_R$ follows from an inequality

$$\|g\|_{C(\Omega)} = \max \left\{ \max_{Q \in \Omega} \left| \iint_\Omega F(Q')K(Q, Q', c) \times \right.\right.$$

$$\times \frac{u(Q')}{\sqrt{u^2(Q') + v^2(Q')}} dQ' \right| \le$$

$$\le \max_{Q \in \Omega} \left| \iint_\Omega F(Q')K(Q, Q', c) \frac{v(Q')}{\sqrt{u^2(Q') + v^2(Q')}} dQ' \right| \right\} \le R,$$

where $f = (u,v)^{\text{T}}$ is any function of the set S_R and $g = Bf \equiv (B_1(u,v),$ $B_2(u,v))^{\text{T}}$. From the last inequality we have also $BS_R \subset S_R$. So, the operator $B = (B_1, B_2)^{\text{T}}$ is completely continuous mapping a closed convex set $S_R \subset C(\Omega)$ into itself.

Theorem is proved.

From the Theorem 1 follows satisfaction of conditions of the Schauder principle [16] according to which the operator $B = (B_1, B_2)^{\text{T}}$ has a fixed point $f_* = (u_*, v_*)^{\text{T}}$ belonging to the set S_R. This point is a solution of a system of Equation (14) and Equation (10), respectively. Substituting $f_* = (u_*, v_*)^{\text{T}}$ into (12), we obtain a solution of (9) being a stationary point of the functional (7).

The solutions of a system of equations analogous with (14) in a case of one-dimensional domains Ω were investigated for the synthesis problem of linear antenna array in particular in [17]. The obtained there results show that for equations of the type (10) and (14) non-uniqueness and branching of solutions dependent on the size of physical parameter are characteristic. Directly the results [17] cannot be transferred on the two-dimensional two-parametric problem (8) and (14). Here, as unlike the points of branching [17], the branching lines of solutions exist and a problem on finding the lines of branching is a nonlinear two-parametrical spectral problem.

Easily to be convinced that function

$$f_0(Q, \mathbf{c}) = \iint_G F(Q')K(Q, Q', \mathbf{c})\,dQ' \tag{18}$$

is one of solutions of (10) in the class of real functions. Since, as shown before, the operator D determined by (13), is positive on the nonnegative functions cone $C(\Omega)$, $DK \subset K$ and $F \subset K$, then $f_0 = DF$ also is a nonnegative function in the domain Ω.

To find the lines of branching and complex solutions of (10), branching-off from real solution $f_0(Q,c)$, we consider a problem on finding such set of values of parameters $c^{(0)} = \left(c_1^{(0)}, c_2^{(0)}\right)$ and all distinct from $f_0(Q,c)$ solutions of the system (14) which for $\left|c - c^{(0)}\right| \to 0$ (where $c_1 \geq c_1^{(0)}, c_2 \geq c_2^{(0)}$) satisfy conditions

$$\max_{Q \in G}\left|\grave{e}\left(Q,c\right) - f\left(Q,c^{(0)}\right)\right| \to 0 \quad \max_{Q \in G}\left|v\left(Q,c\right)\right| \to 0 \tag{19}$$

These conditions indicate the need to find small continuous in G solutions

$$w\left(Q,c\right) = u\left(Q,c\right) - f_0\left(Q,c^{(0)}\right), \quad \omega\left(Q,c\right) = v\left(Q,c\right),$$

which converge uniformly to zero as $c \to c^{(0)}$.

Set

$$c_1 = c_1^{(0)} + \mu, \quad c_2 = c_2^{(0)} + v \tag{20}$$

and desired solutions we find in the form

$$u\left(Q,c\right) = f_0\left(Q,c^{(0)}\right) + w\left(Q,\mu,v\right), \quad v\left(Q,c\right) = \omega\left(Q,\mu,v\right). \tag{21}$$

Further we omit dependence of the functions $w\left(Q,\mu,v\right)$ and $\omega\left(Q,\mu,v\right)$ on parameters μ and v.

Notice the properties of integrand in the system (14). They are continuous functions with respect to the arguments. After substitution (20) and (21) into (14) the integrand develop in equiconvergent power series by functional arguments w and ω, numerical parameters μ and ν in the

vicinity of a point $\left(c^{(0)}, f_0\left(\underline{Q}, c^{(0)}\right), 0\right)$:

$$F(\underline{Q}')K\left(\underline{Q}, \underline{Q}', \mathbf{c}\right)\frac{u(\underline{Q}')}{\sqrt{u^2(\underline{Q}') + v^2(\underline{Q}')}} =$$

$$= \sum_{m+n+p+q\geq 0} A_{mnpq}\left(\underline{Q}, \underline{Q}', \mathbf{c}^{(0)}\right) w^m(\underline{Q}')\omega^n(\underline{Q}')\mu^p\nu^q,$$

$$F(\underline{Q}')K\left(\underline{Q}, \underline{Q}', \mathbf{c}\right)\frac{v(\underline{Q}')}{\sqrt{u^2(\underline{Q}') + v^2(\underline{Q}')}} =$$

$$= \sum_{m+n+p+q\geq 1} B_{mnpq}\left(\underline{Q}, \underline{Q}', \mathbf{c}^{(0)}\right) w^m(\underline{Q}')\omega^n(\underline{Q}')\mu^p\nu^q.$$

$$(22)$$

Here $A_{mnpq}\left(\underline{Q}, \underline{Q}', c^{(0)}\right), B_{mnpq}\left(\underline{Q}, \underline{Q}', c^{(0)}\right)$ are coefficients of expansion continuously dependent on the arguments. Substituting (20) and (22)

into (14) and taking into account that $f_0\left(\underline{Q}, c^{(0)}\right)$ solves the system (14) we obtain a system of nonlinear equations with respect to small solutions w, ω:

$$u(\underline{Q}) = a_{10}\left(\underline{Q}, \mathbf{c}^{(0)}\right)\mu + a_{01}\left(\underline{Q}, \mathbf{c}^{(0)}\right)\nu +$$

$$+ \sum_{m+n+p+q\geq 2} \mu^p\nu^q \iint_{\Omega} A_{mnpq}\left(\underline{Q}, \underline{Q}', \mathbf{c}^{(0)}\right) w^m(\underline{Q}')\omega^n(\underline{Q}')d\underline{Q}',$$

$$(23)$$

$$\omega(Q) - \iint_{\Omega} F(Q)K\left(Q,Q',\mathbf{c}^{(0)}\right)\frac{\omega(Q')}{f_0\left(Q',\mathbf{c}^{(0)}\right)}dQ' =$$

$$= \sum_{m+n+p+q\geq 2} \mu^p \nu^q \iint_{\Omega} B_{mnpq}\left(Q,Q',\mathbf{c}^{(0)}\right)w^m(Q')\omega^n(Q')dQ',$$

$$(24)$$

where

$$a_{10}\left(Q,\mathbf{c}^{(0)}\right) = \iint_{G} A_{0010}\left(Q,Q',\mathbf{c}^{(0)}\right)dQ',$$

$$a_{01}\left(Q,\mathbf{c}^{(0)}\right) = \iint_{G} A_{0001}\left(Q,Q',\mathbf{c}^{(0)}\right)dQ'.$$

NONLINEAR TWO-PARAMETRIC SPECTRAL PROBLEM

For further application of methods of the branching theory of solutions of nonlinear equations [18] to a system (23) and (24) it is necessary to find solutions of distinct from trivial of the linear homogeneous integral equation obtained equating to zero the left part of (24)

$$\varphi(Q) = T(c_1, c_2)\varphi \equiv \iint_{G} \frac{F(Q')}{f_0(Q',c_1,c_2)}K(Q,Q',c_1,c_2)\varphi(Q')dQ'$$

$$(25)$$

under condition $f_0(Q',c) > 0$. Indicate that the operator $T(c):C(\Omega) \to C(\Omega)$ is completely continuous. Proof of this property is similar to the proof of a complete continuity of the operator $B = (B_1, B_2)^\top$ in the Theorem 1.

According to [18] such values of parameters $\left(c_1^{(0)}, c_2^{(0)}\right) \in \mathbb{R}^2$ at which linear homogeneous Equation (25) has distinct from identical zero solutions are points of possible branching of solutions of a system of nonlinear Equations (23) and (24). The eigenfunctions of (25) are used at construction branching-off solutions of (23) and (24).

The spectral parameters c_1 and c_2 are included non- linearly into the kernel of the integral operator. Therefore a problem on finding the distinct from $f_0(Q, c_1, c_2)$ solutions of (25) is a nonlinear two-parametric spectral problem. It consists in finding such values of real parameters $(c_1, c_2) \in \Lambda_c$ at which (25) has distinct from identical zero solutions.

In operational form a nonlinear two-parametric problem is presented as

$$A(c_1, c_2)x \equiv (E - T(c_1, c_2))x = 0.$$

(26)

Here E is an identical operator and $T(c_1, c_2)$ is a linear integrated operator acting in the Banach space $C(\Omega)$. It is necessary to find eigenvalue $c = \left(c_1^{(0)}, c_2^{(0)}\right) \in \Lambda_c$ and corresponding eigenvectors $x^{(0)} \in C(\Omega)$ $\left(x^{(0)} \neq 0\right)$ such that $A\left(c_1^{(0)}, c_2^{(0)}\right)x^{(0)} = 0$.

By direct check we ascertain that for any values of parameters $(c_1, c_2) \in \Lambda_c$ the function

$$\hat{\phi}_0(Q, \mathbf{c}) = \iint_\Omega F(Q')K(Q, Q', \mathbf{c})\, dQ'$$

(27)

is one of eigenfunctions.

Write a conjugate to (25) equation required in later

$$\psi(Q) = T^*(\mathbf{c})\psi \equiv \frac{F(Q)}{f_0(Q, \mathbf{c})} \iint_\Omega K(Q, Q', \mathbf{c})\psi(Q')\, dQ'$$

(28)

At arbitrary $(c_1, c_2) \in \Lambda_c$ the function

$$\hat{\psi}_0(Q) = F(Q).$$

(29)

is one of eigenfunctions of (28)

The existence of distinct from identical zero solutions of (25) at arbitrary $(c_1, c_2) \in \Lambda_c$ testifies to the existence of coherent components of a spectrum conterminous with the domain Λ_c.

For finding the distinct from $\hat{\phi}_0(Q, c)$ solutions we exclude from the kernel of integral Equation (25) the eigen function (27), namely: consider the equation

$$\varphi(Q, \mathbf{c}) = \tilde{T}(\mathbf{c})\varphi \equiv \iint\limits_\Omega \mathrm{K} \ (Q, Q', \mathbf{c}) \varphi(Q') dQ',$$

(30)

where

$$\mathrm{K} \ (Q, Q', \mathbf{c}) = \frac{F(Q')}{f_0(Q', \mathbf{c})} K(Q, Q', \mathbf{c}) - \psi_0(Q)\varphi_0(Q', \mathbf{c})$$

(31)

$$\psi_0(Q) = \frac{\hat{\psi}_0(Q)}{\|\hat{\psi}_0\|_{L_2}}, \quad \varphi_0(Q', \mathbf{c}) = \frac{\hat{\phi}_0(Q', \mathbf{c})}{\|\hat{\phi}_0\|_{L_2}}.$$

(32)

From Schmidt Lemma [18] follows that $\varphi_0(Q, c)$ will not be an eigenfunction of this equation for any values $(c_1, c_2) \in \Lambda_c$. Thus from a spectrum of operator there is excluded coherent component coinciding with the domain Λ_c and corresponding to the function $\varphi_0(Q,c)$.

Using the property of degeneration of the kernel $K(Q,Q',c_1,c_2)$, we reduce (25) to equivalent system of linear algebraic equations having coefficients analytically dependent on parameters c_1, c_2. We write (25) as

$$\varphi(s_1,s_2) = \sum_{n=-N}^{N}\sum_{m=-M}^{M} x_{nm}\exp\left[i(c_1 ns_1 + c_2 ms_2)\right] -$$
$$-x_0\Psi_0(s_1,s_2), \tag{33}$$

where x_{nm}, x_0 are constants determined by the formulas

$$x_{nm} = \frac{c_1}{2\pi}\frac{c_2}{2\pi}\iint_{\Omega}\frac{F(s_1',s_2')}{f_0(s_1',s_2',c_1,c_2)}\exp\left[-i(c_1 ns_1' + c_2 ms_2')\right]\times$$
$$\times\varphi(s_1',s_2')ds_1'ds_2' \quad (n=-N\div N, \ m=-M\div M),$$
$$x_0 = \iint_{\Omega}\varphi_0(s_1',s_2',c_1,c_2)\varphi(s_1',s_2')ds_1'ds_2'.$$

From the Formula (33) follows, that the function $\varphi(S_1,S_2)$ will become known, if will be found x_{nm}, x_0.

Multiplication of both parts of (33) by

$$\frac{F(s_1',s_2')}{f_0(s_1',s_2',c_1,c_2)}\exp\left[-i(c_1 ks_1' + c_2 ls_2')\right]$$

at $k=-N\div N, l=-M\div M$ and by $\varphi_0(S_1',S_2')$, and integration over Ω gives a homogeneous system of the linear algebraic equations for finding x_{nm}, x_0

$$x_{kl} = \sum_{n=-N}^{N} \sum_{m=-M}^{M} a_{nm}^{(kl)}\left(c_1,c_2\right) x_{nm} \quad \begin{pmatrix} k = -N \div N, \\ l = -M \div M \end{pmatrix}.$$

(34)

Here

$$a_{nm}^{(kl)}\left(c_1,c_2\right) = \left\{ t_{nm}^{(kl)}\left(c_1,c_2\right) - \frac{b^{(kl)}\left(c_1,c_2\right)}{1+d_0\left(c_1,c_2\right)} d_{nm}\left(c_1,c_2\right) \right\},$$

$$t_{nm}^{(kl)}\left(c_1,c_2\right) = \frac{c_1 c_2}{4\pi^2} \iint_{\Omega} \frac{F(s_1,s_2)}{f_0(s_1,s_2,c_1,c_2)} \times$$

$$\times \exp\left\{ i\left[c_1\left(n-k\right)s_1 + c_2\left(m-l\right)s_2 \right] \right\} ds_1 ds_2 ,$$

$$b^{(kl)}\left(c_1,c_2\right) = \frac{c_1 c_2}{4\pi^2} \iint_{\Omega} \frac{F(s_1,s_2)}{f_0(s_1,s_2,c_1,c_2)} \psi_0(s_1,s_2) \times ,$$

$$\times \exp\left[-i\left(c_1 k s_1 + c_2 l s_2 \right) \right] ds_1 ds_2 ,$$

$$d_{nm}\left(c_1,c_2\right) = \iint_{\Omega} \varphi_0(s_1,s_2) \exp\left[i\left(c_1 n s_1 + c_2 m s_2 \right) \right] ds_1 ds_2 ,$$

For coefficients of the matrix $A_M\left(c_1,c_2\right) = \left\| \alpha_{nm}^{(kl)}(c_1,c_2) \right\|_{\substack{k,n=-N \div N \\ m,l=-M \div M}}$ the

equality $\alpha_{mn}^{(lk)}(c_1,c_2) = \overline{\alpha_{mn}^{(lk)}(c_1,c_2)}$ is valid, i.e. A_M is the Hermitian or self-adjoint matrix.

Write the equivalent to (26) nonlinear two-paramet— rical spectral problem, corresponding to a system of Equation (34), as

$$A_M(c_1,c_2)x \equiv (E_M - A_M(c_1,c_2))x = 0, \qquad (35)$$

where E_M is a unit matrix of dimension $N_2 \times M_2$.

In order that the system (34) should have distinct from zero solutions, it is necessary

$$\Psi(c_1,c_2) = \det(E_M - A_M(c_1,c_2)) = 0. \qquad (36)$$

It is easy to be convinced, that $\Psi(c_1,c_2)$ is a real function. Really as $T_M(c_1,c_2)$ is the Hermitian matrix then it is obvious that $(E - A_M(c_1,c_2))$ is also the Hermitian matrix. It is known [19] that the determinant of the Hermitian matrix is a real number. So, $\Psi(c_1,c_2)$ is a real function with respect to the real arguments c_1 and c_2 .

Therefore, the problem on finding the set of eigenvalues of (25) or equivalent linear algebraic system (34) is reduced to finding the nulls of the function $\Psi(c_1,c_2)$.

Consider a necessary later on auxiliary one-dimensional spectral problem (as a special case of the problem (35)) on the ray $c_2 = \gamma c_1$ (γ is a real coefficient, $(c_1,c_2) \in \Lambda_c$). Introduce into consideration the matrix - function $\tilde{A}_M(c_1) \equiv A_M(c_1,\gamma c_1)$ and connected with it the one-dimensional spectral problem

$$\tilde{A}_M(c_1,\gamma c_1)x = (E_M - A_M(c_1,\gamma c_1))x = 0. \qquad (37)$$

It is easy to be convinced, that from the properties of coefficients of matrix $A_M(c_1,c_2)$ follows, that the matrix function $A_M(c_1,c_2)$ is continuous and differentiable on the variables in any open and limited domain $\Lambda \subset \Lambda_c \subset \mathbb{R}^2$. In other words $A_M(c_1, c_2)$ is a holomorphic matrix - function, if c_1, c_2 to continue into the domain of complex variables.

Corresponding to (37), Equation (36) has the form

$$\Psi(c_1, \gamma c_1) = \det\left(\mathbf{E}_M - \mathbf{A}_M(c_1, \gamma c_1)\right) = 0 .$$

(38)

We denote the spectrums of the problems (35) and (37) as $s(A)$ and $s(\tilde{A})$, respectively, and the parameter domain c_1 as $\Lambda_{c_1} = \{c_1 : 0 < c_1 \le \alpha\}$. Then for proper ties of the spectrum of (35) the Theorem 1 from [9] is applied which relatively to (35) is formulated thus:

Theorem 2: Let at each $c = (c_1, c_2) \in \Lambda_c$ the matrix $A_M(c_1, c_2) \in L\left(\mathbf{C}^{N_2 \times M_2}, \mathbf{C}^{N_2 \times M_2}\right)$ be the Fredholm operator with a zero index, the matrix - function $A(\bullet, \bullet) : \Lambda_c \to L\left(\mathbf{C}^{N_2 \times M_2}, \mathbf{C}^{N_2 \times M_2}\right)$ be holomorphic in the domain Λ_c and $s(\tilde{A}) \ne \Lambda_{c_1}$. Moreover, let function $\psi(c_1, c_2)$ be continuously differentiable in Λ_c. Then:

1. Each point of a spectrum $c_1^{(0)} \in s(\tilde{A})$ is isolated and it is eigenvalue of the matrix - function $\tilde{A}(c_1) \equiv A(c_1, \gamma c_1)$, to it is corresponding the finite- dimensional eigensubspace $N\left(\tilde{A}\left(c^{(0)}_1\right)\right)$ and finite- dimensional root subspace;

2. Each point $c^{(0)} = \left(c_1^{(0)}, \gamma c_1^{(0)}\right) \in \Lambda_c$ is a point of spectrum of the matrix - function $A(\lambda_1, \lambda_2)$;

3. If $\psi'_{c_2}\left(c^{(0)}_1, c^{(0)}_2\right) \neq 0$ then in some vicinity of the point $c^{(0)}_1$ there is a unique continuous differentiable function $c_2 = c_2(c_1)$ solving the Equation (36), i.e. in some bicircular domain $\Lambda_0 = \left\{ (c_1, c_2) : \left| c_1 - c^{(0)}_1 \right| < \varepsilon_1, \left| c_2 - c^{(0)}_2 \right| < \varepsilon_2 \right\}$ there exists a connected component of spectrum of the matrix-function $A(c_1, c_2)$ (where ε_1, ε_2 are small real constants).

Proof of this theorem concerning the nonlinear two-parametrical spectral problem of the type (35) for more general case (when the operators E and $T(c_1, c_2)$ act in the infinite dimensional Banach space) is presented in [9]. For satisfaction of conditions of Theorem 1 from [9] it is necessary to show that the matrix - function $A(c_1, c_2)$ is the Fredholm matrix at $(c_1, c_2) \in \Lambda_c$. This property follows from the known equality [19] dim(kerA) = dim(kerA *) .

The existence of connected components of spectrum of the matrix - function $A(c_1, c_2)$, under condition of $\Psi'_{c_2}\left(C^{(0)}_1, C^{(0)}_1\right) \neq 0$, follows from the existence theorem of implicitly given function [20, 21].

Let $c^{(i)}_1$ be a root of (38). Then $\left(c^{(i)}_1, c^{(i)}_2 = \gamma c^{(i)}_1\right) \in \Lambda_c$ is eigenvalue of the problem (33). Consider the equation $\Psi(C_1, C_2) = 0$ as a problem on finding the implicitly given function $c_2 = c_2(c_1)$ in the vicinity of a point $c^{(i)}_1$ for which the conditions of existence theorem [21] are satisfied. Hence we have the Cauchy problem

$$\frac{dc_2}{dc_1} = -\frac{\Psi'_{c_1}(c_1, c_2)}{\Psi'_{c_2}(c_1, c_2)},$$

$$(39)$$

$$c^{(i)}_2\left(c^{(i)}_1\right) = \gamma c^{(i)}_1.$$

$$(40)$$

Solving numerically (39) and (40) in some vicinity of a point $c_1^{(i)}$, we find the i -th connected component of spectrum (spectral line) of the matrix - function $A_M(c_1, c_2)$.

By found solutions of the Cauchy problem at the fixed values $\left(c_1^{(i)}, c_2^{(i)}\right)$ the eigenfunctions of (25) are determined through the eigenvectors of

the matrix $A_M\left(c_1^{(i)}, c_2^{(i)}\right)$ obtained by the known methods. Thus four-dimensional matrix A_M is reduced to two-dimensional one by means of corresponding renumbering of elements.

NUMERICAL ALGORITHM OF FINDING THE SOLUTIONS OF A NONLINEAR EQUATION

Show one of iterative processes for numerical finding the solutions of the system (14) based on the successive approximations method [2]:

$$u_{n+1}(Q) = B_1(u_n, v_n) \equiv \iint_\Omega K(Q, Q', \mathbf{c}) F(Q') \times$$

$$\times \frac{u_n(Q')}{\sqrt{u_n^2(Q') + v_n^2(Q')}} dQ',$$

$$v_{n+1}(Q) = B_2(u_n, v_n) \equiv \iint_\Omega K(Q, Q', \mathbf{c}) F(Q') \times$$

$$\times \frac{v_n(Q')}{\sqrt{u_n^2(Q') + v_n^2(Q')}} dQ' \quad (n = 0, 1, ...).$$

$$(41)$$

After substituting the function arg $f_n(Q) = \operatorname{arctg}\left(v_n(Q)' u_n(Q)\right)$ (obtained on the basis of successive approximations (41)) into (12), we denote the obtained sequence of function values as $\{I_n\}$. For the sequence $\{I_n\}$ the Theorem 4.2.1 from [3] is fulfilled. From here follows, that the sequence $\{I_n\}$ is a relaxation one for the functional (7) and numerical sequence $\{\sigma(I_n)\}$ is convergent.

At realization of the iterative process (41) in the case of even on both arguments function $F(s_1, s_2)$ and symmetric domains G and Ω it is expedient to use the property of invariance of integral operators $B_1(u,v)$, $B_2(u,v)$ in the system (14) concerning the type of parity of functions $u(s_1,s_2)$, $v(s_1,s_2)$. The functions u,v having certain type of evenness on corresponding arguments belong to the appropriate invariant sets U_{ij} , $V_{k\ell}$ of the space $c(\Omega)$. Here the indices i,j,k,ℓ have values 0 or 1. In particular, if $u(s_1,s_2) \in U_{01}$ then $u(-s_1,s_2) = u(s_1,s_2)$ and $u(s_1,-s_2) = -u(s_1,s_2)$. By direct check we are convinced that such inclusions take place:

$$B_1(U_{ij} \cup V_{k\ell}) \subset U_{ij}, \qquad B_2(U_{ij} \cup V_{k\ell}) \subset V_{k\ell},$$
$$\mathbf{B}(U_{ij} \cup V_{k\ell}) \subset U_{ij} \cup V_{k\ell}.$$

The possibility of existence of fixed points of the operator B belonging to appropriate invariant set (i.e. solutions of system (14) and, respectively, Equation (10)) follows from these relations.

NUMERICAL EXAMPLE

Consider an example of approximation of the function $F(s_1,s_2) = \cos(\pi s_1 / 2)|\sin(\pi s_2)|$ (Figure 1), given in the domain $G = \{(s_1,s_2) : |s_1| \le 1, |s_2| \le 1\} \subset G\Omega$, for $N_2 \times M_2 = 11 \times 11$ and values of parameters $c_1 = 1.6$ and $c_2 = 1.2$ belonging to the ray $c_2 = 0.75c_1$. The possible branching lines of solutions of the system (14) and accordingly the Equation (10), as solutions of two-dimensional spectral problem (25), are shown in Figure 2. Here the first branching lines are denoted by numbers 1 and 2. To the solutions branching-off at the points of these lines there correspond the odd on s_2 functions arg f (s_1,

s_2) and the coefficients of transformation $I_{n,m}$ $(n = -N \div N, m = -M \div M)$ are real, but nonsymmetrical concerning to the plane XOZ .

In Figure 3 in logarithmic scale are presented values of the functional σ obtained on the solutions of two types at values of parameter c_2 $=0.75c_1$: the curve 1 corresponds to solutions in a class of real functions $f_0 = (Q)$, curve 2 – to the branching-off solution with odd on s_2 argument arg $f(s_1, s_2)$.

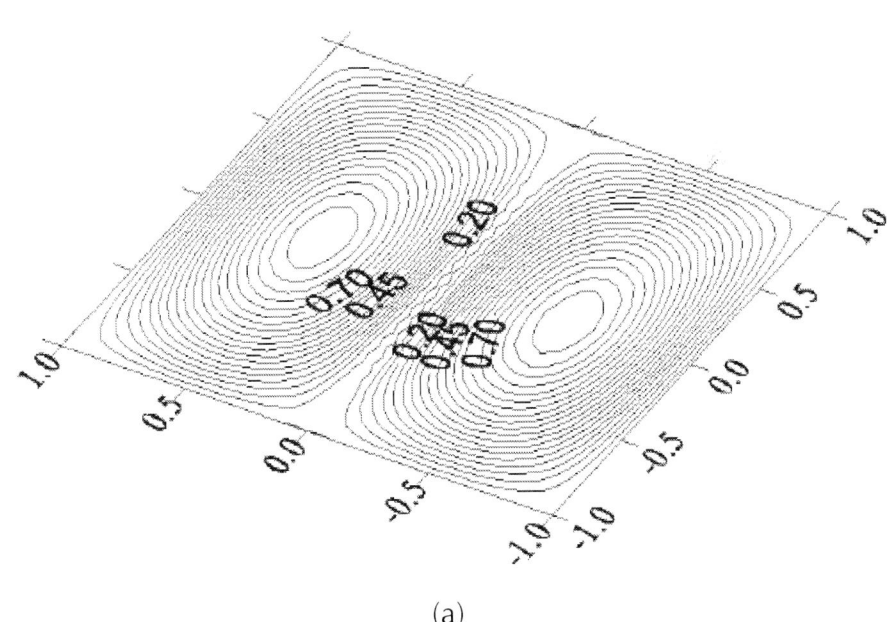

(a)

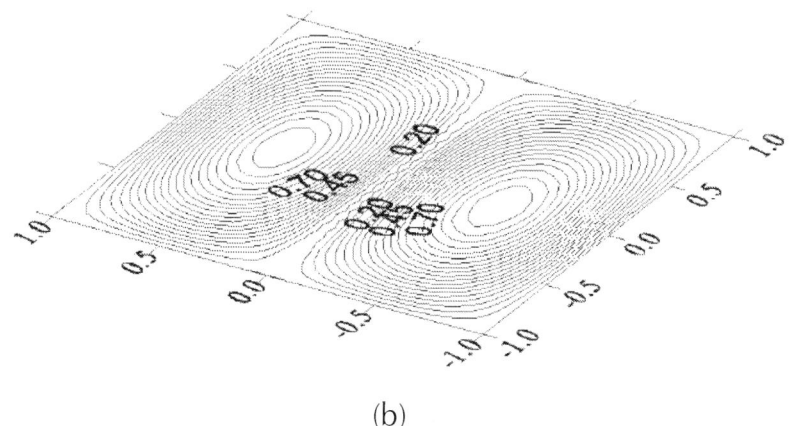

(b)

Figure 1: The function $F(s_1, s_2) = \cos(\pi s_1 / 2)\left|\sin(\pi s_2)\right|$ given in the domain $\overline{G} = \{(s_1, s_2) : |s_1| \le 1, |s_2|\} \subset \Omega.$

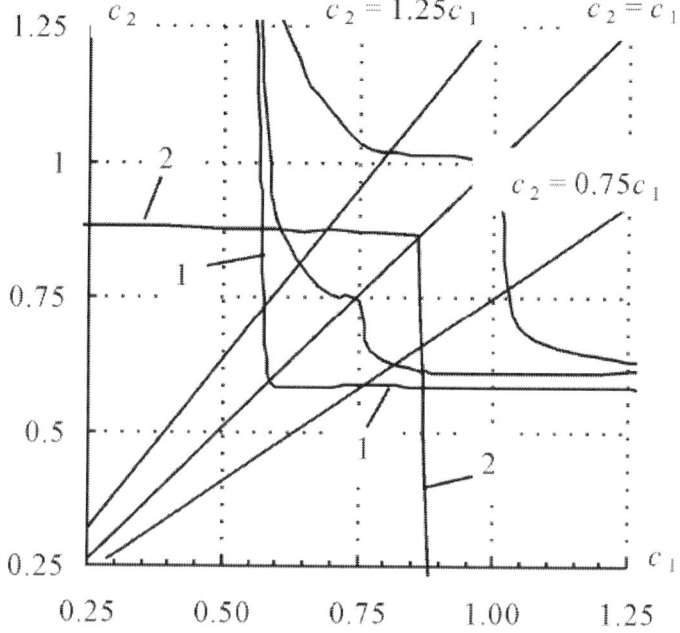

Figure 2: The branching lines of solutions.

From analysis of Figure 3 follows that at the point $c_1 0.77$ from real solution branch-off more effective complex-conjugate between themselves solutions, on which the functional s accepts smaller values, than on the real solution. If to introduce into consideration parameter $C_2 = Mc_2$ characterizing the quantity of basic functions in transformation (1), the identical efficiency of approximation (identical values of the functional s on real and branching-off solutions) is reached with use of the branching-off solution at decrease of the quantity of basic functions on the value $C_2 = 0.75Dc1$.

An amplitude (**a**) and argument (**b**) of approximate function are given in Figure 4 for $c_1 = 1.6$ and $c_2 = 1.2$. The amplitude values of the Fourier Transform coefficients corresponding to this solution are shown in Figure 5. As we see in figure, the values of amplitudes of coefficients are nonsymmetrical concerning the plane YOZ , but the amplitude of approximate function (Figure 4, **a**) is symmetric.

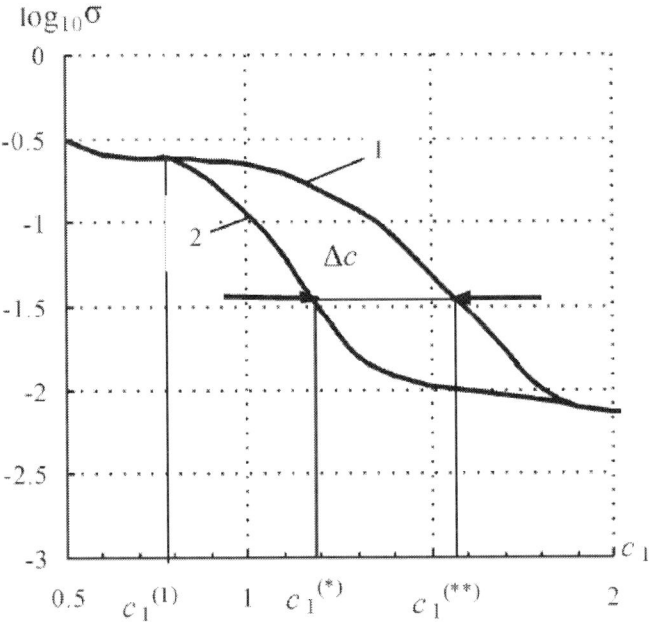

Figure 3: The values of functional on initial and branching- off solutions.

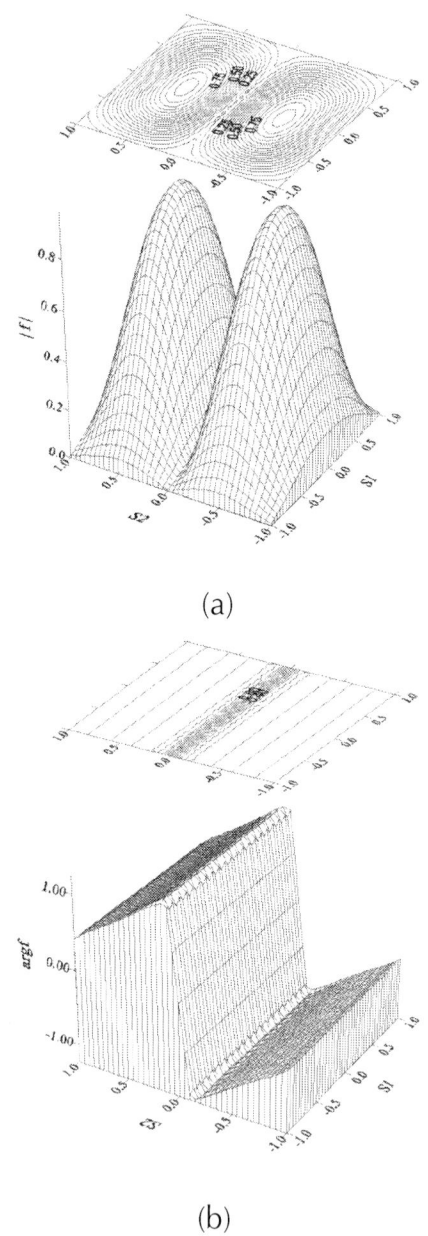

(a)

(b)

Figure 4: The modulus (**a**) and argument (b) of approximation function.

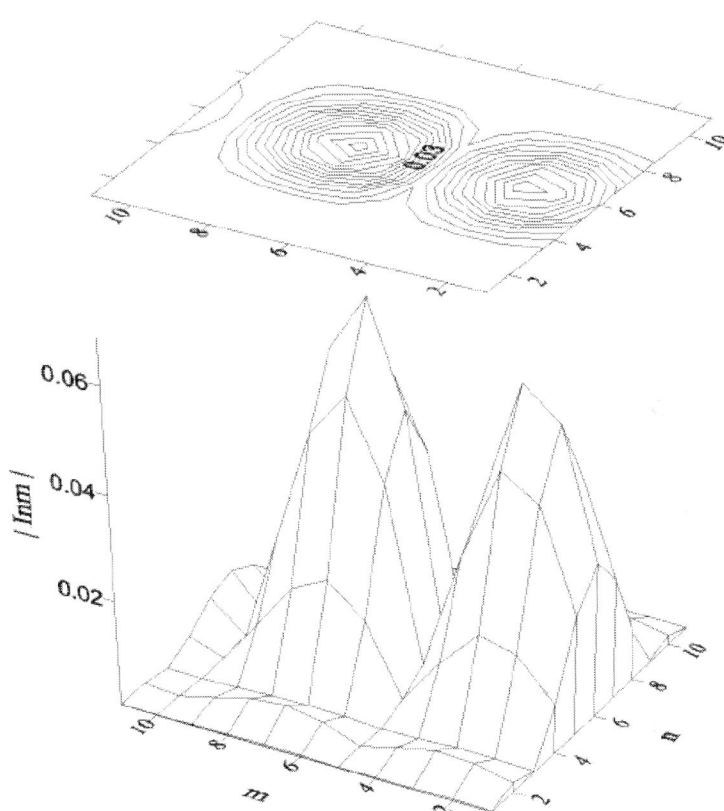

Figure 5: The optimum amplitude of Fouier transform coefficients.

For comparison of approximate functions, corresponding to different solutions of (10), the curves corresponding to different types of the presented solutions in the section $s_1 0$ are given in Figure 6. The curve 1 corresponds to the given function $F(0, s_2)$, the curve 2 – to branching-off solution, the curve 3 – to real solution $f_0 (0, s_2)$. Obviously that the branching-off solution better (in meaning of the functional s) approaches the prescribed function by the module.

CONCLUSIONS

Mark the basic features and problems arising at investigation of the considered class of tasks:

The basic difficulty to solve this class of problems is study of nonuniqueness and branching of existing solutions dependent on the parameters c_1, c_2 entering into the discrete Fourier Transform.

As follows from investigations, presented, in particular, in [3,17] (for a special case, when $F(s_1, s_2) = F_1(s_1) \times F_2(s_2)$), the quantity of the existing solutions grows considerably with increase of the parameters c_1, c_2. Let us indicate, that in many practical applications, in particular, in the synthesis problems of radiating systems, it is important to obtain the best approximation to the given function $F(s_1, s_2)$ at rather small values of parameters c_1, c_2. This allows limiting by investigation of several first points (lines) of branching.

To find the branching points (lines) of solutions of (8), it is necessary, as opposed to [3, 17], to solve not enough studied multiparametric spectral problem. The offered in this work approaches allow to find the solutions of a nonlinear two-parametric spectral problem for homogeneous integral equations with degenerate kernels analytically dependent on two spectral parameters.

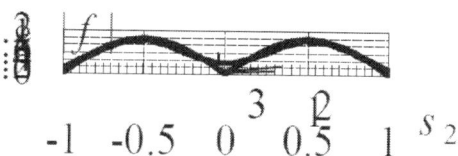

Figure 6: The given (curve 1) and approximation functions in the section $s_2 = 0$ corresponding to branching-off (curve 2) and real (curve 3) solutions.

When finding the solutions to a system of Equation (14) by successive approximations method, to obtain the solutions of a certain type of parity of the function arg f (s_1, s_2) it is necessary to choose an initial approximation ( ) arg $f_0 s_1, s_2$ of the same type of parity according to (42).

To obtain the irrefragable answer concerning the branching-off solutions for certain values of parameters c_1, c_2 it is necessary to use the branching theory of solutions [18]. It is the object of special investigations.

REFERENCES

1. B. M. Minkovich and V. P. Jakovlev, "Theory of Synthesis of Antennas, " Soviet Radio, Moscow, 1969.
2. P. A. Savenko, "Numerical Solution of a Class of Nonlinear Problems in Synthesis of Radiating Systems," Computational Mathematics and Mathematical Physics, Vol. 40, No. 6, 2000, pp. 889-899.
3. P. O. Savenko, "Nonlinear Problems of Radiating Systems Synthesis (Theory and Methods of the Solution)," Institute for Applied Problems in Mechanics and Mathematics, Lviv, 2002.
4. G. M. Vainikko, "Analysis of Discretized Methods," Tartus Gos. University of Tartu, Tartu, 1976.
5. R. D. Gregorieff and H. Jeggle, "Approximation von Eigevwertproblemen bei nichtlinearer Parameterabh?ngi- keit," Manuscript Math, Vol. 10, No. 3, 1973, pp. 245- 271.
6. O. Karma, "Approximation in Eigenvalue Problems for Holomorphic Fredholm Operator Functions I," Numerical Functional Analysis and Optimization, Vol. 17, No. 3-4, 1996, pp. 365-387.
7. M. A. Aslanian and S. V. Kartyshev, "Updating of One Numerous Method of Solution of a Nonlinear Spectral Problem," Journal of Computational Mathematics and Mathe- matical Physics, Vol. 37, No. 5, 1998, pp. 713-717.
8. S. I. Solov'yev, "Preconditioned Iterative Methods for a Class of Nonlinear Eigenvalue Problems," Linear Algebra and its Applications, Vol. 41, No. 1, 2006, pp. 210-229.
9. P. A. Savenko and L. P. Protsakh, "Implicit Function Method in Solving a Two-dimensional Nonlinear Spectral Problem," Russian Mathematics (Izv. VUZ), Vol. 51, No. 11, 2007, pp. 40-43.
10. V. A. Trenogin, "Functional Analysis," Nauka, Moscow ,1980.
11. I. I. Privalov, "Introduction to the Theory of Functions of Complex Variables," Nauka, Moscow, 1984.
12. A. N. Kolmogorov and S. V. Fomin, "Elements of Functions Theory and Functional Analysis," Nauka, Moscow, 1968.

13. P. P. Zabreiko, A. I. Koshelev and M. A. Krasnoselskii, "Integral Equations," Nauka, Moscow, 1968.
14. M. A. Krasnoselskii, G. M. Vainikko, and P. P. Zabreiko, "Approximate Solution of Operational Equations," Nauka, Moscow, 1969.
15. I. I. Liashko, V. F. Yemelianow and A. K. Boyarchuk, "Bases of Classical and Modern Mathematical Analysis," Vysshaya Shkola Publishres, Kyiv, 1988.
16. E. Zeidler, "Nonlinear Functional Analysis and Its Appli- cations I: Fixed-Points Theorem," Springer-Verlag, New York, Berlin, Heidelberg, Tokyo, 1985.
17. P. A. Savenko, "Synthesis of Linear Antenna Arrays by Given Amplitude Directivity Pattern," Izv. Vysch. uch. zaved. Radiophysics, Vol. 22, No. 12, 1979, pp. 1498-1504.
18. M. M. Vainberg and V. A. Trenogin, "Theory of Branching of Solutions of Nonlinear Equations," Nauka, Moscow, 1969.
19. V. V. Voyevodin and Y. J. Kuznetsov, "Matrices and Calcu- lations," Nauka, Moscow, 1984.
20. A. Gursa, "Course of Mathematical Analysis, Vol. 1, Part 1," Moscow-Leningrad, Gos. Technical Theory Izdat, 1933.
21. V. I. Smirnov, "Course of High Mathematics, Vol. 1," Nauka, Moscow, 1965.

CITATION

P. Savenko and M. Tkach, "Numerical Approximation of Real Finite Nonnegative Function by the Modulus of Discrete Fourier Transform," Applied Mathematics, Vol. 1 No. 1, 2010, pp. 65-75. doi: 10.4236/am.2010.11008.

The Role Of Asymptotic Mean In The Geometric Theory Of Asymptotic Expansions In The Real Domain

Antonio Granata
Department of Mathematics and Computer Science, University of Calabria, Rende (Cosenza), Italy

ABSTRACT

We call "asymptotic mean" $(at +\infty)$ of a real-valued function $f \in L^1_{loc}[T, +\infty)$ the number, supposed to exist, $M_f := \lim_{x \to +\infty} X^{-1} \int_T^x f(t)dt$, and highlight its role in the geometric theory of asymptotic expansions in the real domain of type (*) $f(X) = a_1\phi_1(x) + L + a_n\phi_n(x) + o(\phi_n(x)), x \to +\infty$, where the comparison functions $\phi_1(x), L, \phi_n(x)$, forming an asymptotic scale at $+\infty$, belong to one of the three classes having a definite "type of variation" at $+\infty$, slow, regular or rapid. For regularly- varying comparison functions we can characterize the existence of an asymptotic expansion (*) by the nice property that a certain quantity $F(t)$ has an asymptotic mean at $+\infty$. This quantity is defined via a linear differential operator in f and admits of a remarkable geometric interpretation as it measures the ordinate of the point wherein that special curve $y = a_1(t)\phi_1(x) + L + a_n(t)\phi_n(x)$, which has a contact of order n − 1 with the graph of f at the generic point t, intersects a fixed vertical line, say x = T. Sufficient or necessary conditions hold true for the other two classes. In this article we give results for two types of expansions already studied in our current development of a general theory of asymptotic expansions in the real domain, namely polynomial and two-term expansions.

INTRODUCTION

In our current endeavor to establish a general analytic theory of asymptotic expansions in the real domain [1] -[6] , we highlighted that what we called the geometric approach leads in a natural way to a linear differential operator, say F , depending solely on the comparison functions appearing in a possible expansion; certain asymptotic or integral conditions involving the quantity $F[f(x)]$ then characterize an expansion of a given function f either in itself or matched to other expansions obtained by formal differentiation in suitable senses. The theory we are referring to is based on the following ideas. Suppose one wishes to find conditions (sufficient and/or necessary) for the validity of an asymptotic expansion

$$f(x) = a_1\phi_1(x) + \cdots + a_n\phi_n(x) + o(\phi_n(x)), \ x \to +\infty,$$

(1.1)

where the ordered n-tuple of comparison functions $(\phi_1(x),\cdots,\phi_n(x))$ forms an asymptotic scale at $+\infty$, that is to say: $\phi_i(x) >> \phi_{i+1}(x)$ i.e. $\phi_{i+1}(x) = o(\phi_i(x)), x \to +\infty; i = 1,\cdots,n-1$. In this paper we intentionally choose $x \to +\infty$ as this is the situation wherein the classical concept of asymptotic mean plays a role. The simplest elementary case is that of an "asymptotic straight line"— $f(x) = ax + b + o(1), x \to \infty,$,—and it goes back to Newton the "natural" idea of looking at this contingency as the "limit position of the tangent line at the graph of f" as the point of tangency goes to infinity. The German geometer Haupt [7] , in 1922, extended this idea to study "nth-order asymptotic parabolas" i.e. "polynomial asymptotic expansions"

$$f(x) = a_nx^n + \cdots + a_1x + a_0 + o(1) \equiv P_n(x) + o(1), \ x \to +\infty,$$

(1.2)

looking at them as "limit positions of nth-order osculating parabolas". In [1] we collected various scattered results on such expansions completing them with some missing links and adding a new theory called "factorizational theory". A rich bibliography with historical references is also to be found in [1] . For a general expansion (1.1) a rough idea consists in looking at the "generalized polynomial"

$\sum_{i=1}^{n} a_i \phi_i(x)$ as the limit position of a suitable family of "generalized polynomial curves"

$$y = a_1(t)\phi_1(x) + \cdots + a_n(t)\phi_n(x),$$ (1.3)

as the parameter $t \to +\infty$. Of course a curve (1.3) must have some meaningful link with the graph of f and, from a technical point of view, the simplest choice consists in (1.3) admitting of a contact of order $(n-1)$ with $y = f(x)$ at the generic point $(t, f(t))$, i.e.

$$\sum_{i=1}^{n} a_i(t)\phi_i^{(k)}(t) = f^{(k)}(t), 0 \le k \le n-1.$$ (1.4)

This requires suitable assumptions: the regularity of the ϕ_i's and f and a special structure of the n-tuple $\phi_1(x), L, \phi_n(x)$. Then the theory consists in characterizing the contingency

$$\lim_{t \to +\infty} a_i(t) \equiv \gamma_i \in \mathbb{R},$$ (1.5)

via a certain set of asymptotic relations for f. At least this is what has been done for the two cases already systematized in the literature: that of polynomial asymptotic expansions in [1] and that of two-term expansions in [4] . In this paper we point out that, whenever the comparison functions admit of an "index of variation at $+\infty$", one can obtain new types of asymptotic results revolving around a classical concept which we label "asymptotic mean". In §2 we first present an overview of the class of functions with an asymptotic mean; then, after introducing classes of slowly-varying, regularly-varying or rapidly-varying functions in a restricted sense, we give new results correlating these last classes, asymptotic means and weighted asymptotic means.

In §3 we give characterizations of certain sets of polynomial asymptotic expansions via asymptotic means of the coefficients of nth-order osculating parabolas; in particular we shall study the following

Conjecture. An asymptotic expansion (1.2) holds true iff the constant coefficient of the nth-order osculating parabola at the generic point $(t, f(t))$ has an asymptotic mean at $+\infty$.

This nice statement will be proved true for a class of functions f satisfying a certain differential inequality. In §4 we establish either characterizations or sufficient conditions or necessary conditions for an asymptotic expansion

$$f(x) = a_1 \phi_1(x) + a_2 \phi_2(x) + o(\phi_2(x)), \; x \to +\infty,$$

$$(1.6)$$

according to the three "types of variation at $+\infty$" of the comparison functions ϕ_i so giving the exact results vaguely mentioned in ([4] ; pp. 261-263).

Extension of the results to a general asymptotic expansion (1.1), n ≥ 3, is based on information about the asymptotic behavior of Wronskians of regularly- or rapidly-varying functions and this requires a separate non- short treatment.

Almost all proofs are collected in §5. A recurrent notation is:

• $f \in AC^0(I) \equiv AC(I) \Leftrightarrow AC(I)$ is absolutely continuous on each compact interval of I;

• $f \in AC^k(I) \Leftrightarrow f^{(k)} \in AC(I)$.

FUNCTIONS WITH AN ASYMPTOTIC MEAN

General Properties

The following concept is meaningful in itself and often encountered both in classical Analysis (see references throughout this section) and in modern applied mathematics, Sanders and Verhulst [8] .

Definition 2.1. If $f \in L^1_{loc}[T,+\infty]$ then its asymptotic mean at $+\infty$ is defined as the number

$$M_f := \lim_{x \to +\infty} \frac{1}{x} \int_1^x f(t)dt$$

(2.1)

provided that the limit exists and is finite. (Obviously neither the existence nor the value of M_f depend on the particular choice of T.)

We shall use the symbol M_∞ to denote the class of all functions defined on an interval of the form $[T,+\infty)$ and having an asymptotic mean at $+\infty$; $M\infty$ is obviously a vector space over $\mathbb{R}$. In order to help the reader grasp the meaning of the quantity M_f we shall list various classes of functions contained in M_∞; at the same time we shall have at our disposal some practical rules for testing the existence and the possible value of M_f.

1. If $f(+\infty)$ exists in the extended real line (for instance if f is monotonic) then

    $f \in M_\infty$ iff $f(+\infty) \in \mathbb{R}$: in such a case $M_f = f(+\infty)$. Just apply L'Hospital's

    rule to the quotient $\int_T^x f(t)dt \Big/ x$.

2. If f is periodic on $[T,+\infty)$ with period $p \neq 0$ then

    $$M_f = \frac{1}{p} \int_T^{T+p} f(t)dt \equiv \text{the arithmetic mean of } f \text{ on } [T, T+p].$$

    (2.2)

A direct elementary proof may be found in Corduneanu ([9] ; Remark, p. 24).

3. If f is almost periodic on $\mathbb{R}$ then $f \in M_\infty$, see ([9] ; pp. 23-24). This property is essential to develop a theory of Fourier series for almost-periodic functions.

4.   If f has a bounded antiderivative (i.e. $\sup_{x\in[T,+\infty)}\left|\int_T^x f\right| < +\infty$) then $M_f = 0$. This is the condition appearing in the classical Dirichlet test for convergence of improper integrals of type $\int^{+\infty} f\phi$. If, in particular, the improper integral $\int^{+\infty} f$ converges then $M_f = 0$.

5.   If $\int^{+\infty} |f|^p < +\infty$ for some p, $1 \le p < +\infty$, then $M_f = 0$. This follows from the previous case when p = 1 and from Hölder's inequality, when $p > 1$:

$$\int_T^x f(t)\,dt \equiv \int_T^x t^\alpha \left[t^{-\alpha} f(t) \right] dt = o\left(x^\alpha \right),\ x \to +\infty,$$

(2.3)

6.   If the improper integral $\int^{+\infty} t^{-a} f(t)\,dt$ converges for some a, $0 < a \le 1$, then $M_f = 0$. The proof is an immediate consequence of the relation

$$\int_T^x f(t)\,dt \equiv \int_T^x t^\alpha \left[t^{-\alpha} f(t) \right] dt = o\left(x^\alpha \right),\ x \to +\infty,$$

(2.4)

which follows from the hypothesis and the next

Proposition 2.1. If $\int^{+\infty} f$ converges then for any $a > 0$:

$$\int_T^x t^\alpha f(t)\,dt = o\left(x^\alpha \right),\ x \to +\infty.$$

(2.5)

In fact integrating by parts we have

$$\int_T^x t^\alpha f(t)\,dt = -\int_T^x t^\alpha d\left(\int_t^{+\infty} f \right) = c - x^\alpha \left(\int_x^{+\infty} f \right) + \alpha \int_T^x t^{\alpha-1} \left(\int_t^{+\infty} f \right) dt,$$

(2.6)

where $c = T^\alpha \left(\int_T^{+\infty} f \right)$. That the last term on the right is $o\left(x^\alpha \right)$ follows dividing by x^α and applying l'Hospital's rule.

Proposition 2.1 is widely used in asymptotic theory of ordinary differential equations: in a different but equivalent formulation it goes back to Faedo ([10] ; lemma, p. 118) and also appears in a paper by

Hallam ([11] ; lemma 1.1, p. 136). However the nontrivial proofs given by these authors are only valid for one-signed f. The elementary proof given above applies to any f: it essentially goes back to Hukuhara ([12] ; Lemma 1, p. 72) and appears again in Ostrowski ([13] ; Lemma II).

7.   If for some fixed $\lambda > 0$ there exists a finite limit

$$\lim_{x \to +\infty} e^x \int_x^{+\infty} f(t) e^{-t} dt \equiv L,$$

(2.7)

then $f \in M_\infty$ and $M_f = \frac{a}{\lambda}$. For a proof see Agnew ([14] ; Th. 6.2, p. 17).

8.   If there exists a finite limit

$$\lim_{x \to +\infty} e^x \int_x^{+\infty} f(t) e^{-t} dt \equiv L,$$

(2.8)

then $f \in M_\infty$ and $M_f = L$. This has been proved by Agnew ([14] ; Th. 4.2, p. 13) using a non-elementary indirect argument based on the foregoing result and another theorem of his.

9.   If $f \in AC[T, +\infty)$ it is a trivial fact that relation

$$f(x) = ax + o(x), \quad x \to +\infty,$$

(2.9)

does not necessarily imply $f'(+\infty) = a$, the converse inference being true; but relation (2.9) is equivalent to $f' \in M_\infty$ and, if this is the case, then $a = M_{f'}$. In fact

$$\frac{1}{x} \int_T^x f' = \frac{f(x) - f(T)}{x} = \frac{f(x)}{x} + o(1), \quad x \to +\infty.$$

(2.10)

The last relation also implies the following version of L'Hospital's rule for functions in M_∞

240

Introduction to Real Analysis

Proposition 2.2: If $f,g \in AC[T,+\infty), g(x) \neq 0 \forall x, f'$ and $g' \in M_\infty$ and $Mg' \neq 0$ then the $\lim x_{\to+\infty} \dfrac{f(x)}{g(x)}$ exists in $\mathbb{R}$ and equals $\dfrac{M_{f'}}{M_{g'}}$.

For the proof just write $\dfrac{f(x)}{g(x)} = \dfrac{f(x)}{x}\dfrac{x}{g(x)}$, and apply (2.10).

10. The space M_∞ has a link with the classical concept of Cesàro-summability.

A function $f \in L^1_{loc}[T,+\infty)$ is said to be Cesàro-summable of order one, or summable $(C,1)$, on $[T,+\infty)$ if the following limit

$$\lim_{x\to+\infty}\frac{1}{t}\int_T^t\left(\int_T^\tau f\right)d\tau \quad \textit{exists as a finite number.}$$

(2.11)

This concept is an extension to improper integrals of the concept of arithmetical mean for a sequence, see Hardy ([15] ; pp. 430-434) and ([16] ; Ch. V and p. 110). It follows from our definition that "f is summable (C, 1) on $[T,+\infty)$ iff $F(t) \equiv \int_T^t f \in M_\infty$".

11. Two negative properties concerning functions in M_∞.

a. Not any bounded function belongs to M_∞. Counterexample:

$$f(x) := \sin(\log x) + \cos(\log x); \quad \int_1^x f = x\sin(\log x),$$

(2.12)

even if f is uniformly continuous on $[1,+\infty)$. In Blinov [17] there is a more elaborate counterexample of a bounded uniformly-continuous function constructed with the implicit use of almost-periodic functions.

For f bounded, the contingency "$f \in M_\infty$" can be characterized via the behavior at the origin of the Laplace-transform of f: see either Ditkine and Proudnikov ([18] ; Th. 4, p. 196) or Baumgärtel and Wollenberg

([19] ; Ch. 6, pp. 97-98) where the problem is treated in a functional-analytic context.

a. In general no information on the order of growth of a function in M_∞ can be drawn. For the function

$$f(x) = 1 + (\alpha + 2) x^{\alpha+1} \cos\left(x^{\alpha+2}\right), \alpha + 2 > 0, \tag{2.13}$$

we have

$$\int_0^x f = x + \sin\left(x^{\alpha+2}\right) = x + O(1), x \to +\infty \quad \left(\text{hence } M_f = 1\right), \tag{2.14}$$

But $\overline{\lim}_{x \to +\infty} \dfrac{f(x)}{x^a} = +\infty$; $\underline{\lim}_{x \to +\infty} \dfrac{f(x)}{x^a} = -\infty$. All the above properties, from 1 to 9, practically are sufficient conditions for $f \in M_\infty$, none of them being characteristic. A counterexample for the converse of property in 6 is provided by:

$$\begin{cases} f(x) := 1/\log x; \ \int_2^x 1/\log t \, dt \sim x/\log x, x \to +\infty \left(\text{hence } M_f = 0\right) \\ \int_x^{+\infty} \left(t^\alpha \log t\right)^{-1} dt \text{ diverges for each } \alpha, 0 \leq \alpha \leq 1. \end{cases} \tag{2.15}$$

12. However in Ostrowski ([20] ; IV, pp. 65-68) the following characterization is reported:

13. The number M_f in Definition 2.1 exists iff

$$\exists \lim_{x \to +\infty} x \int_x^{+\infty} t^{-2} f(t) \, dt \equiv L_f \in \mathbb{R}, \tag{2.16}$$

and, if this is the case, $M_f = L_f$. This result, used by Ostrowski, e.g., in the study of Frullani's integral, may also yield the nice geometric characterization of a rectilinear asymptote, see (3.15) below. But in other asymptotic investigations a more general form of condition (2.16) is encountered, namely

$$\exists \lim_{x \to +\infty} \phi(x) \int_x^{+\infty} \left(1/\phi(t)\right)' f(t)\,dt \equiv L_{f,\phi} \in \mathbb{R},$$

(2.17)

where ϕ stands for some suitable function such that $\phi(+\infty) = +\infty$. The number $L_{f,\phi}$ is a kind of "weighted asymptotic mean" of f and can be considered, the sign apart, as a "generalized limit of $f(x)$ as $x \to +\infty$" for the simple reason that a trivial application of L'Hospital's rule yields $\lim_{x \to \infty} \dfrac{\int_x^{+\infty} \left(\frac{1}{\phi(t)}\right)' f(t)dt}{\left(\frac{1}{\phi(x)}\right)} = -\lim_{x \to \infty} f(x),$ under obvious hypotheses on f, ϕ. The notion of regular variation gives the key to finding out a large meaningful class of test-functions ϕ, including powers, such that (2.17), valid for one fixed ϕ, is equivalent to $f \in M_\infty$.

Preliminaries On Regularly- Or Rapidly-Varying Functions

We use the notion of variation, either regular or rapid, in a restricted sense; for the general theory the reader is referred to the monograph by Bingham, Goldie and Teugels [21] . We get three different results for the three classes defined in

Definition 2.2: Let $\phi \in AC[T, +\infty), \phi(x) > 0$ for each x large enough.
I. ϕ is termed "regularly varying at $+\infty$ (in the strong sense)" if

$$\phi'(x)/\phi(x) = \alpha x^{-1} + o\left(x^{-1}\right), x \to +\infty,$$

(2.18)

for some constant $a \in \mathbb{R}$ which is called the index of regular variation of ϕ at $+\infty$. The family of all such functions for a fixed a is denoted by $\mathcal{R}_a(+\infty)$. In the case $a = 0$ the function ϕ is also termed "slowly varying at $+\infty$ (in the strong sense)".
II. ϕ is termed "rapidly varying at $+\infty$ (in the strong sense)" if

$$\lim_{x \to +\infty} x\phi'(x)/\phi(x) = \pm\infty.$$

(2.19)

Accordingly, the index of rapid variation at $+\infty$ is defined to be either $+\infty$ or $-\infty$ and the corresponding families of functions are denoted by $\mathcal{R}_{+\infty}(+\infty)$ and $\mathcal{R}_{-\infty}(+\infty)$.

III. ϕ is said to have an "index of variation at $+\infty$ in the strong sense" if the following limit exists in the extended real line:

$$\lim_{x \to +\infty} x\phi'(x)/\phi(x) \equiv \alpha, \quad -\infty \le \alpha \le +\infty. \tag{2.20}$$

Remarks 1: Condition "ϕ ultimately of one strict sign" is essential both in the general and in our restricted definition. The choice $\phi > 0$ is merely conventional. Writing $|\phi| \in \mathcal{R}_a(+\infty)$ tacitly implies "$\phi \in AC[T, +\infty)$ for some T and $(\phi) \ne 0$ for x large enough".

2. Typical functions in $\mathcal{R}_a(+\infty), a \in \mathcal{R}$, are: $x^a \cdot \prod_{k=1}^{n}(1_k(x))^{\beta_k}$ where 1^k denotes the k-time iterated logarithm, $1_1 \equiv \log$, and α, β_k's are any real numbers. Typical functions in $\mathcal{R}_a(+\infty)$ are: $x^a \cdot \left[\prod_{k=1}^{n}(1^k(x))^{\beta_k}\right] \cdot \exp(cx^y)(\alpha, \beta_k \in \mathcal{R}; c \ne 0, y > 0)$ Here the index of variation is: $(\text{sign } c) \cdot \infty$.

3. For $\alpha \ne 0, \phi'$ too has ultimately one strict sign and there are two contingencies for the limit

$$\lim_{x \to +\infty} \phi(x) = \begin{cases} 0 & \text{if } -\infty \le \alpha < 0, \\ +\infty & \text{if } 0 < \alpha \le +\infty, \end{cases} \tag{2.21}$$

as inferrred from the identity $\log \phi(x) = c + \int_T^x [\phi'(t)/\phi(t)] dt.$

For $a = 0$ all the possible contingencies may occur for this limit as shown by the functions: 1; $(\log(x))^{\alpha}$, with $\alpha \neq 0$; $2 + \sin\left[(\log(x)^{\alpha}\right]$, $0 < \alpha < 1$.

4.   If $\phi \in AC^{1}[T, +\infty) \cap \mathcal{R}_{a}(+\infty)$ with $-\infty \leq \alpha \leq \pm\infty$, it may happen that $|\phi'|$ has no index of variation at $+\infty$ as shown by the counterexamples:

$$\begin{cases} x^{\alpha} + \sin x \in \mathcal{R}_{\alpha}(+\infty), \ 1 < \alpha \leq 2, & (x\phi''/\phi' \text{ oscillatory and unbounded for } \alpha < \\ 2 + \sin\left[(\log x)^{\alpha}\right] \in \mathcal{R}_{0}(+\infty), \ 0 < \alpha < 1, & (\phi' \text{ oscillatory}); \\ 2e^{x} + \sin(e^{x}) \in \mathcal{R}_{+\infty}(+\infty), & (x\phi''/\phi' \text{ oscillatory and unbounded}); \\ e^{-x} + e^{-2x}\sin(e^{x/2}) \in \mathcal{R}_{-\infty}(+\infty), & (x\phi''/\phi' \text{ oscillatory and unbounded}). \end{cases}$$

$$(2.22)$$

But if $|\phi'|$ has an index of variation then there are precise links between the two indexes.

Lemma 2.3: If $\phi \in AC^{1}[T, +\infty)$ and if both ϕ and $|\phi'|$ have indexes of variation at $+\infty$, respectively α and α', then:

$$\begin{cases} \alpha' = \alpha - 1 \text{ if } \alpha \in \mathbb{R} \setminus \{0\} \text{ or if } \alpha = 0 \text{ and } \lim_{x \to +\infty} \phi(x) = \text{ either } 0 \text{ or } +\infty; \\ \alpha' = \alpha \text{ if } \alpha = \pm\infty. \end{cases}$$

$$(2.23)$$

In the case $a = 0$ and without the stated additional condition on $\lim_{x \to +\infty} \phi(x)$, it may happen that $|\phi'| \in \mathcal{R}_{\alpha'}(+\infty)$ with $-\infty \leq \alpha' \leq -1$ as shown by the simple examples:

$$1 + e^{-x}; \ 1 + x^{-\delta} \ (\delta > 0); \ 1 + (\log x)^{-\delta} \ (\delta > 0);$$

$$(2.24)$$

but it cannot be $\alpha' > -1$.

Relationships Between Asymptotic Mean And Weighted Asymptotic Means

We can now give and understand generalizations of the mentioned results by Ostrowski and Agnew.

Theorem 2.4: Let $\phi \in AC^1[T,+\infty)$ and $\phi(+\infty) = +\infty$.

I.   (Regularly-varying functions: extension of a result by Ostrowski, 1976). If

$$\phi \in \mathcal{R}_\alpha(+\infty) \text{ for some real number } \alpha > 0, |\phi'| \in \mathcal{R}_{\alpha-1}(+\infty),$$

$$(2.25)$$

then for any fixed $f \in L^1_{loc}[T,+\infty)$ conditions (2.1) and (2.17) are equivalent to each other. If this is the case then $M_f = -L_{f,\phi}$, hence $L_{f,\phi}$ does not depend on ϕ. An equivalent statement is:

Under conditions (2.25) the following two asymptotic relations are equivalent to each other:

$$\int_x^{+\infty} (1/\phi(t))' f(t)\,dt = \frac{a}{\phi(x)} + o(1/\phi(x)), \quad \int_T^x f(t)\,dt = -ax + o(x), \ x \to +\infty,$$

$$(2.26)$$

for a constant a which turns out to depend only on f. In one direction we have that the first relation in (2.26), which is trivially true whenever $f(+\infty) = -\alpha$, holds true under the weaker condition $f \in M_\infty$.

II.   Slowly-varying functions). If

$$\phi \in \mathcal{R}_0(+\infty), |\phi'| \in \mathcal{R}_{-1}(+\infty),$$

$$(2.27)$$

then for any fixed $f \in L^1_{loc}[T,+\infty)$ condition (2.1) implies (2.17) with $M_f = -L_{f,\phi}$.

III. Rapidly-varying functions: extension of a result by Agnew, 1942). If

$$\phi'(x) \neq 0 \quad \text{for } x \text{ large enough,}$$

$$(2.28)1$$

$$\phi(x)/\phi'(x) = o(x), \quad x \to +\infty,$$

$$(2.28)2$$

$$\left(\phi(x)/\phi'(x)\right)' = o(1), \quad x \to +\infty,$$

$$(2.28)3$$

(which imply that both $\phi, |\phi'|$ are rapidly-varying at $+\infty$) then for any fixed $f \in L^1_{loc}[T, +\infty)$ condition (2.17) implies (2.1) with $M_f = -L_{f,\phi}$.

Corollary 2.5. Special cases reformulated:

$$\int_x^{+\infty} t^{-\alpha} f(t)\,dt = L x^{1-\alpha} + o\left(x^{1-\alpha}\right), x \to +\infty \Leftrightarrow M_f = (\alpha-1)L, \ (\alpha > 1);$$

$$(2.29)$$

$$f \in M_\infty \Rightarrow \int_x^{+\infty} (\log t)^{-\beta} t^{-1} f(t)\,dt = \frac{M_f}{\beta-1}(\log x)^{1-\beta} + o\left((\log x)^{1-\beta}\right), x \to +\infty, (\beta > 1),$$

$$(2.30)$$

$$\begin{cases} \int_x^{+\infty} t^{\gamma-1} \exp\left(-ct^\gamma\right) f(t)\,dt = \frac{L}{c\gamma} \exp\left(-ct^\gamma\right) + o\left(\exp\left(-cx^\gamma\right)\right), x \to +\infty, (c, \gamma > 0) \\ \Rightarrow f \in M_\gamma \text{ and } M_f = L. \end{cases}$$

$$(2.31)$$

For $\alpha = 2$ the equivalence in (2.29) is Ostrowski's result, see (2.16), and for $c = y = 1$ the inference in (2.31) is Agnew's result, see (2.9). A counterexample for the converse inference in part (II) is provided by:

$$\begin{cases} \phi_1(x) := (\log x)^{\alpha-1} \ (\alpha > 1), \ f(x) := \sin(\log x) + \cos(\log x); \\ \phi \in R_0(+\infty), \phi' \in R_{-1}(+\infty); \ f \notin M_\infty \text{ as } \int_1^x f = x \sin(\log x); \\ \int_x^{+\infty} t^{-1} (\log t)^{-\alpha} f(t)\,dt = o\left((\log x)^{1-\alpha}\right), x \to +\infty; \end{cases}$$

$$(2.32)$$

where the last relation can be easily proved by suitably integrating by parts.

And a counterexample for the converse inference in part (III) is trivially provided by:

$$\begin{cases} \phi_2(x) := e^x, \ f(x) := \sin x; \ M_f = 0; \\ e^x \int_x^{+\infty} e^{-t} \sin t\,dt = \frac{1}{2}(\sin x + \cos x) \quad \text{admits of no limit as } x \to +\infty. \end{cases}$$
(2.33)

Notice that $\phi, |\phi'|$ may be rapidly varying without satisfying $(2.28)_3$ as shown by the function $\phi_3(x) := \exp(2x + \sin x)$. We do not know if part (III) remains true when replacing the three conditions (2.28) by the weaker conditions: $\phi, |\phi'| \in \mathcal{R}_{\pm\infty}(+\infty)$.

We add the following isolated result, needed in the sequel, without placing it in a general context.

Proposition 2.6: If $f \in M_\infty$ then:

$$g(x) := x^{-\alpha} f(x) \in M_\infty \quad \text{and} \quad M_g = 0 \ \forall \alpha > 0.$$
(2.34)

We end this section by mentioning that the concept of asymptotic mean plays a role also in "Tauberian theorems", Hardy ([16] ; Ch. 12), in non-oscillation properties of second-order differential equations, Hartman [22] and ([23] ; pp. 365-367), and in the theory of Cauchy-Frullani integrals, Ostrowski [20] . In this last paper our Theorem 2.4-(I) appears for the first time in the literature though for the special case $\phi(x) = x$ and the proof is somewhat involved. In a previous paper Ostrowski ([13] ; Lemma II) had given a quick proof of a lemma correlated to our present context, a proof based on integration by parts; curiously enough he does not apply the same elementary device in proving the result under consideration, which is just the device used by us to prove the general case. Also the original proof by Agnew [14] is indirect; the author is interested in studying the limit

$$\lim_{x \to +\infty} \int_x^{x+\lambda} f(t) \, dt = \lambda M_f \quad \text{for each fixed } \lambda \in \mathbb{R}$$

(2.35)

where M_f is a real number independent from λ. He first proves the equivalence between (2.8) and (2.35) and then that (2.35) implies $f \in M_\infty$.

POLYNOMIAL ASYMPTOTIC EXPANSIONS AND ASYMPTOTIC MEANS

If $f \in AC^{n-1}[T, +\infty), n \geq 1$, then $f^{(n)}(t)$ is defined almost everywhere and for each such t let us consider the "nth-order osculating parabola" to the graph of f at the point $(t, f(t))$:

$$y = f(t) + (x-t)f'(t) + \frac{(x-t)^2}{2!} f''(t) + \cdots + \frac{(x-t)^n}{n!} f^{(n)}(t),$$

(3.1)

which may be rewritten in the form

$$y = \sum_{k=0}^n F_{n,k}(t) x^k \equiv P_n(x;t),$$

(3.2)

where $P_n(x;t)$ is a polynomial in x of degree $\leq n$, whose coefficients $F_{n,k}(t)$ depend on the parameter t. If all the limits

$$\lim_{t \to +\infty} F_{n,k}(t) \equiv \gamma_{n,k} \quad \left(k = 0,1,\cdots,n \text{ and } t \text{ running through the set where } f^{(n)} \text{ is defined}\right)$$

(3.3)

exist as finite numbers, we say that the parabola

$$y = \gamma_{n,n} x^n + \cdots + \gamma_{n,1} x + \gamma_{n,0} \equiv \Pi_n(x),$$

(3.4)

or equivalently the polynomial $\prod_n(x)$, is the "nth-order limit parabola" to [the graph of] f at $+\infty$. A limit parabola of order zero denotes a mere relation $f(x) = a_0 + o(1), x \to +\infty$

We shall call the function $F_{n,0}(t)$ the "nth-order contact indicatrix" of the curve $y = f(x)$ with respect to the y-axis as it represents the ordinate of the point of intersection in the x, y-plane between the y-axis and the curve (3.1).

We report here simplified versions of two of the main results in [1].

Proposition 3.1: For $f \in AC^{n-1}[T, +\infty), n \geq 1$, the following are equivalent properties:

1. The graph of f has a limit parabola at $+\infty$ of order n i.e., by definition, all the limits

$$F_{n,k}(+\infty) \equiv \gamma_{n,k}, (k = 0, 1, \cdots, n), \text{ exist as numbers.}$$

(3.5)

2. The single limit

$$F_{n,0}(+\infty) \equiv \gamma_{n,0} \text{ exists as a finite number.}$$

(3.6)

3. There exists a polynomial $\prod_n(x) := \sum_{h=0}^{n} \gamma_{n,h} x^h$ such that

$$f^{(k)}(x) = \prod_n^{(k)}(x) + o(x^{-k}), x \to +\infty; k = 0, 1, \cdots, n; \prod_n^{(0)} \equiv \prod_n.$$

(3.7)

If this is the case then the following integral representation holds true

$$f(x) = \prod_n(x) + n! x \int_x^{+\infty} dt_1 \cdots \int_{t_{n-2}}^{+\infty} dt_{n-1} \int_{t_{n-1}}^{+\infty} t^{-n-1} \left[F_{n,0}(t) - \gamma_{n,0} \right] dt, x \geq T > 0,$$

(3.8)

for a suitable polynomial $\prod_n$, the same as above, and a suitable number $\gamma_{n,0}$, the same as in (3.6).

We expressed relations in (3.7) by saying that the asymptotic expansion

$$f(x) = \Pi_n(x) + o(1), \quad x \to +\infty,$$
(3.9)

is formally differentiable n times in the "strong sense" because in the same paper we characterized another weaker set of differentiated expansions, ([1] ; Th. 3.1, p. 173), which we shall not presently use.

Proposition 3.2: If $f \in AC^{n-1}[T,+\infty)$ and is convex of order $n \geq 1$ on

$[T,+\infty)$—which is equivalent to the property that $(-1)^n F_{n,0}$ is increasing thereon—then: f has a "polynomial asymptotic expansion at $+\infty$", i.e. it satisfies a relation of type

$$f(x) = a_n x^n + \cdots + a_1 x + a_0 + o(1) \equiv P_n(x) + o(1), \quad x \to +\infty,$$
(3.10)

iff its nth-order contact indicatrix $F_{n,0}$ is bounded (hence, by monotonicity, condition (3.6) holds true). If this is the case then we also have the properties in Proposition 3.1, hence the expansion (3.10) automatically implies its formal differentiability n times in the strong sense.

Now we give analogues of the two foregoing propositions with condition (3.6) replaced by the weaker condition $F_{n,0} \in M_\infty$; strong differentiability will be granted $(n-1)$ times and the validity of an expansion (3.10) will be characterized for a class of functions larger than nth-order convexity.

Theorem 3.3: For $f \in AC^{n-1}[T,+\infty), n \geq 1$, the following are equivalent properties:

1. All the functions

$F_{n,k} \in M_\infty$; $M_{n,k} :=$ *the asymptotic mean at* $+\infty$ *of* $F_{n,k}$; $0 \leq k \leq n$.
(3.11)

2. The single function

$$F_{n,0} \in M_\infty.$$
(3.12)

3. There exists a polynomial $P_n(x) := \sum_{k=0}^n a_k x^k$ such that

$$f^{(k)}(x) = P_n^{(k)}(x) + o(x^{-k}), \quad x \to +\infty; \; k = 0,1,\cdots,n-1.$$

$$(3.13)$$

If this is the case then $a_k = M_{n,k}, \forall k$ and the following integral representation holds true:

$$f(x) = M_{n,n}x^n + L + M_{n,1}x + n!x \int_x^{+\infty} dt_1 \, L \int_{t_{n-2}}^{+\infty} dt_{n-1} \int_{t_{n-1}}^{+\infty} t^{-n-1} F_{n,0}(t) dt, \; x \ge T > 0.$$

$$(3.14)$$

In the elementary case n = 1 the result is:

$$\begin{cases} f(x) = ax + b + o(1), \; x \to +\infty \iff F_{1,0} \equiv -xf'(x) + f(x) \in \mathcal{M}_\infty; \\ and \; in \; this \; case : \; a = M_{f'}, b = M_{1,0}. \end{cases}$$

$$(3.15)$$

Notice that the representation of $f^{(n)}$ inferred from (3.14) contains the quantity $x^{-n}F_{n,0}(x)$ hence, by the example in (2.13), no information on the growth-order of $f^{(n)}$ may be obtained in the context of Theorem 3.3, generally speaking.

For $n \ge 2$ a characterization similar to that in (3.15) holds true under a restriction on the sign of $F_{n,0}$ and we have the following analogue of Proposition 3.2.

Theorem 3.4: Let $f \in AC^{n-1}[T,+\infty), n \ge 2$ and let $F_{n,0}$ satisfy a one-sided boundedness condition:

$$either \quad F_{n,0}(x) \le c \; \forall \, x \ge T \quad or \quad c \le F_{n,0}(x) \; \forall \, x \ge T.$$

$$(3.16)$$

Then an expansion (3.10) holds true iff $F_{n,0} \in \mathcal{M}_\infty$. If this is the case then, according to Theorem 3.3, the expansion (3.10) is formally differentiable $(n-1)$ times in the strong sense.

We exhibit an example for the case n = 1 and a counterexample for the case n = 2; they seem to be just the same because in both expansions the remainder is exactly the same quantity but a striking difference appears in the behaviors of $F_{1,0}$ and $F_{2,0}$.

Example for the case $n = 1$:

$$
\begin{cases}
f_1(x) := ax + b - \int_x^{+\infty} dt \int_T^t \left(\tau^{-1}\sin\tau\right)' d\tau, \quad (T > 0 \text{ and such that } \sin T = 0); \\
f_1(x) = ax + b + o(1); \quad f_1'(x) = a + \int_T^x \left(t^{-1}\sin t\right)' dt = a + x^{-1}\sin x = a + O\left(x^{-1}\right); \\
f_1^{(k)}(x) = O\left(x^{-1}\right) \quad \forall k \geq 2; \quad F_{1,0}(x) \equiv b - \sin x + o(1).
\end{cases}
$$

$$(3.17)$$

Here $F_{1,0}$ is bounded and admits of asymptotic mean ($= b$) but has no limit at $+\infty$; accordingly the expansion $f_1(x) = ax + b + o(1)$ is not formally differentiable in the strong sense though the differentiated expansions of any order satisfy the remarkable asymptotic estimates in (3.17).

Counterexample for the case $n = 2$:

$$
\begin{cases}
f_2(x) := ax^2 + bx + c - \int_x^{+\infty} dt \int_T^t \left(\tau^{-1}\sin\tau\right)' d\tau, \quad (T > 0 \text{ and such that } \sin T = 0); \\
f_2(x) = ax^2 + bx + c + o(1); \quad f_2'(x) = 2ax + b + x^{-1}\sin x = 2ax + b + O\left(x^{-1}\right); \\
f_2''(x) = 2a - x^{-2}\sin x + x^{-1}\cos x = 2a + O\left(x^{-1}\right); \quad f_2^{(k)}(x) = O\left(x^{-1}\right) \quad \forall k \geq 3; \\
F_{2,0}(x) \equiv \frac{x^2}{2} f_2''(x) - xf_2'(x) + f_2(x) = c - \frac{3}{2}\sin x + \frac{1}{2}x\cos x - \int_x^{+\infty} dt \int_T^t \left(\tau^{-1}\sin\tau\right)' d\tau = \frac{1}{2}x\cos x + O(1)
\end{cases}
$$

$$(3.18)$$

Here $F_{2,0}$ is unbounded both from below and from above and admits of no asymptotic mean; notwithstanding, an asymptotic expansion $f_2(x) = ax^2 + bx + c + o(1)$ holds true. Hence the equivalence stated in Theorem 3.4 may fail without the restriction in (3.16). According to Theorem 3.3 the expansion of f_2 is not formally differentiable once in the strong sense.

In the elementary case in (3.15) condition $F_{1,0} = o(1,)x \to +\infty$ is explicitly defined in Giblin ([24] ; p. 279) as the "bounded distance condition" and it is easily checked that it is equivalent to a pair of relations

$$f(x) = ax + O(1), \; f'(x) = a + O(x^{-1}), \; x \to +\infty;$$

(3.19)

it is the further condition of existence of asymptotic mean that changes the first relation in (3.19) into an asymptotic straight line.

TWO-TERM ASYMPTOTIC EXPANSIONS AND ASYMPTOTIC MEANS

In this section we give an exhaustive list of results concerning the role of asymptotic mean in the theory of two-term asymptotic expansions involving comparison functions admitting of indexes of variation at $+\infty$. We first report a result from [4] .

Preliminary notations and formulas ([4] ; p. 255). As usual we say that two functions f, g (as well as their graphs) have a first-order contact at a point t_0 if $f(t_0) = g(t_0)$ and $f'(t_0) = g'(t_0)$ provided that f, g are defined on a neighborhood of t_0 and the involved derivatives exist as finite numbers.

Let now ϕ_1, ϕ_2 be two real-valued functions differentiable on an interval I such that their Wronskian $W(x) := W(\phi_1(x), \phi_2(x))$ never vanishes on I and let f be differentiable on I. Then for each $t_0 \in I$ there exists a unique function in the family $F := \mathrm{span}(\phi_1, \phi_2)$ having a first-order contact with f at t_0. Denoting this function by $F^*(x; t_0)$ we have

$$F^*(x; t_0) = f_1^*(t_0)\phi_1(x) + f_2^*(t_0)\phi_2(x), \quad x \in I,$$

(4.1)

Where

$$\begin{cases} f_1^*(t_0) := W\big(f(t_0),\phi_2(t_0)\big)/W(t_0) = \big(f(t)/\phi_2(t)\big)'\big/\big(\phi_1(t)/\phi_2(t)\big)' & \text{evaluated at } t = t_0, \\ f_2^*(t_0) := -W\big(f(t_0),\phi_1(t_0)\big)/W(t_0) = \big(f(t)/\phi_1(t)\big)'\big/\big(\phi_2(t)/\phi_1(t)\big)' & \text{evaluated at } t = t_0. \end{cases}$$

$$(4.2)$$

If $f \in F$ then $F^*(x;t_0) = f*(x)$ on I for any chosen t_0. The function

$$F^*(t) := F^*(T;t) \equiv \phi_1(T)f_1^*(t) + \phi_2(T)f_2^*(t) \quad t \in I,$$

$$(4.3)$$

will be called the contact indicatrix of order one of the function f at the point t with respect to the family f and the straight line $x = T$.
$f*(t)$ represents the ordinate of the point of intersection between the vertical line $x = T$ and the curve $y = f^*{}_1(t)\phi_1(x) + f^*{}_2(t)(x)$ where t is thought of as fixed. The assumption on $W(x)$ implies that ϕ_1 and ϕ_2 do not vanish simultaneously hence F^* is a nontrivial linear combination of f_1*, f_2*. It may happen that, for some choices of T, F^* coincides with f_1* or f_2*, a constant factor apart, according as $\phi_2(T) = 0$ or $\phi_1(T) = 0$.
Using (4.2) F^* may be represented as

$$\begin{aligned} F^*(x) &= \frac{1}{W(x)}\Big[\phi_1(T)W\big(f(x),\phi_2(x)\big) - \phi_2(T)W\big(f(x),\phi_1(x)\big)\Big] \\ &= \frac{1}{W(x)} \cdot W\big(f(x),\phi_1(T)\phi_2(x) - \phi_2(T)\phi_1(x)\big) \\ &\equiv W\big(\Phi(x),f(x)\big)/W(x), \end{aligned}$$

$$(4.4)$$

where we have put

$$\Phi(x) := \phi_2(T)\phi_1(x) - \phi_1(T)\phi_2(x).$$

$$(4.5)$$

Proposition 4.1. (Characterization of a two-term asymptotic expansion: [4] , Th. 4.4, p. 258). Assumptions:

$$\phi_1, \phi_2 \in C^1\left[T, x_0\right[, T \in \mathbb{R}; \quad \phi_2(x) = o\big(\phi_1(x)\big), x \to x_0^-;$$

(4.6)

$$\phi_1(x), \phi_2(x), W(x) := W\big(\phi_1(x), \phi_2(x)\big) \neq 0 \quad \forall x \in \left[T, x_0\right[.$$

(4.7)

For a function $f \in L^1_{loc}[, T \in \mathbb{R};$ the following are equivalent properties:
1. It holds true an asymptotic expansion

$$f(x) = a_1 \phi_1(x) + a_2 \phi_2(x) + o\big(\phi_2(x)\big), \quad x \to x_0^-.$$

(4.8)

2. There exists a finite limit

$$\lim_{x \to x_0^-} \frac{\phi_1(x)}{\phi_2(x)} \cdot \int_x^{x_0} \left(\frac{\phi_2(t)}{\phi_1(t)}\right)' f_2^*(t) dt = -m.$$

(4.9)

3. There exists a finite limit

$$\lim_{x \to x_0^-} \frac{\Phi(x)}{\phi_2(x)} \cdot \int_x^{x_0} \left(\frac{\phi_2(t)}{\Phi(t)}\right)' F^*(t) dt = -\ell.$$

(4.10)

If this is the case we have the following two representations:

$$f(x) = a_1 \phi_1(x) + a_2 \phi_2(x) - \phi_1(x) \cdot \int_x^{x_0} \left(\frac{\phi_2(t)}{\phi_1(t)}\right)' \left[f_2^*(t) - m\right] dt, \quad x \in [T, x_0[;$$

(4.11)

$$f(x) = a_1 \phi_1(x) + a_2 \phi_2(x) - \Phi(x) \cdot \int_x^{x_0} \left(\frac{\phi_2(t)}{\Phi(t)}\right)' \left[F^*(t) - \ell\right] dt, \quad x \in]T, x_0[.$$

(4.12)

The validity of (4.8) may be expressed by the geometric locution: "the graph of f admits of the curve $y = a_1\phi_1(x) + a_2\phi_2(x)$ as an asymptotic curve in the family $F \equiv \text{span}(\phi_1, \phi_2)$, as $x \to x_0^-$."

Notice that in the cited reference condition (4.10) is written in the form

$$\lim_{x \to x_0^-} \frac{\Phi(x)}{\phi_2(x)} \cdot \int_x^{x_0} W(t)(\Phi(t))^{-2} F^*(t) dt \equiv -\frac{\ell}{\phi_2(T)};$$

(4.13)

however (4.5) implies

$$\left(\phi_2/\Phi\right)' \equiv W(\Phi, \phi_2) \cdot \Phi^{-2} = \phi_2(T) \cdot W(\phi_1, \phi_2) \cdot \Phi^{-2},$$

(4.14)

and (4.10) follows.

The two limits in (4.9), (4.10) are of the type studied in §2 and a direct application of Theorem 2.4 gives the following results.

Theorem 4.2: In assumptions (4.6)-(4.7) let it be: $x_0 = +\infty$;

$\phi_1 / \phi_2 \in AC^1[T, +\infty); f \in L^1_{loc}[T, +\infty)$

I. (Regularly-varying comparison functions). If

$$\phi_1/\phi_2 \in \mathcal{R}_\alpha(+\infty) \text{ for some real number } \alpha > 0, \left|(\phi_1/\phi_2)'\right| \in \mathcal{R}_{\alpha-1}(+\infty),$$

(4.15)

then the following three properties are equivalent:

$$f(x) = a_1\phi_1(x) + a_2\phi_2(x) + o(\phi_2(x)), \ x \to +\infty \ \text{(with suitable constants } a_i \text{)};$$

(4.16)

$$f_2^* \in \mathcal{M}_\infty;$$

(4.17)

$$F^* \in \mathcal{M}_\infty.$$

(4.18)

II. (Slowly-varying comparison functions). If

$$\phi_1/\phi_2 \in \mathcal{R}_0(+\infty), \quad \left|(\phi_1/\phi_2)'\right| \in \mathcal{R}_{-1}(+\infty),$$ (4.19)

then each condition (4.17) or (4.18) implies an expansion (4.16).

III. (Rapidly-varying comparison functions). Put $\phi^\%. = \phi_1\phi_2$ and suppose that:

$$\begin{cases} \tilde{\phi}(x)/\tilde{\phi}'(x) = o(x), x \to +\infty, \\ \left(\tilde{\phi}(x)/\tilde{\phi}'(x)\right)' = o(1), x \to +\infty; \end{cases}$$ (4.20)

then an expansion (4.16) implies both conditions (4.17)-(4.18).
Under the stated assumptions for the validity of part (I) the equivalence
"(4.16) Û (4.18)" admits of the following geometric reformulation:
"The graph of f admits of an asymptotic curve in the family

$F \equiv \text{span}(\phi_1,\phi_2)$, as $x \to +\infty$, iff the contact indicatrix of order one of
the function f with respect to F has an asymptotic mean at $+\infty$".
Notice that this result for two-term expansions requires no restrictions

on the signs of f^*_2, F^*.

PROOFS

Proof of Lemma 2.3. By hypothesis the following two limits exist in $\overline{\mathbb{R}}$:

$$\lim_{x \to +\infty} x\phi'(x)/\phi(x) \equiv \alpha,$$

$$\lim_{x \to +\infty} x\phi''(x)/\phi'(x) \equiv \alpha'.$$ (5.1)

We now evaluate α by L'Hospital's rule first noticing that: $0 < a \leq +\infty$
implies $\phi(+\infty) = +\infty$, whereas for $-\infty < a < 0$ it is $\phi(+\infty) = 0$ and the

first limit in (5.1) implies $\lim_{x \to +\infty} x\phi'(x) = 0$. In both cases the rule may be applied and

$$\alpha = \lim_{x \to +\infty} \frac{\phi'(x) + x\phi''(x)}{\phi'(x)} = 1 + \alpha'.$$

(5.2)

It remains the case $\alpha' = -\infty$ which implies $\phi(+\infty) = 0$ and this condition leads to excluding the following contingencies for the indicated reasons:

1. $\alpha' = +\infty \Rightarrow |\phi'(+\infty)| = +\infty \Rightarrow \phi(+\infty) = +\infty (\text{being } \phi > 0)$. Type equation here.

2. $-1 < \alpha' < +\infty \Rightarrow x|\phi'(x)| \in \mathcal{R}_{a'+1}(+\infty) \Rightarrow \lim_{x \to +\infty} xx|\phi'(x)| = +\infty \Rightarrow$
   (by L'Hospital's rule)

$$\Rightarrow \lim_{x \to +\infty} \left| \left(\int_T^x \phi' \right) \middle/ x\phi'(x) \right| = \lim_{x \to +\infty} \left| \frac{\phi'(x)}{\phi'(x) + x\phi''(x)} \right| = \lim_{x \to +\infty} \left| \frac{1}{1 + \frac{\phi''(x)}{\phi'(x)}} \right| = \frac{1}{|1 + a'|},$$

which is a positive real number; hence $\int_T^{+\infty} |\phi'| = +\infty$ which would imply $\phi(+\infty) = +\infty$.

3. $-\infty < a' < -1 \Rightarrow x|\phi'(x)| \in \mathcal{R}_{a'+1}(+\infty) \Rightarrow \lim_{x \to +\infty} x|\phi'(x)| = 0$, and this would imply, by L'Hospital's rule as in (5.2):

$$-\infty = \lim_{x \to +\infty} x \frac{\phi'(x)}{\phi(x)} = 1 + a'.$$

4. The case $a' = -1$ must be treated in a different way. A basic property of our class of functions, directly inferred from the limits in (5.1), claims the validity of the following asymptotic estimates:

$$\begin{cases} \phi \in \mathcal{R}_\alpha(+\infty), \alpha \in \mathbb{R} \Rightarrow x^{\alpha-\epsilon} \ll \phi(x) \ll x^{\alpha+\epsilon}, x \to +\infty, \forall \epsilon > 0; \\ \phi \in \mathcal{R}_{-\infty}(+\infty) \Rightarrow \phi(x) \ll x^{-\alpha}, x \to +\infty, \forall \alpha > 0; \\ \phi \in \mathcal{R}_{+\infty}(+\infty) \Rightarrow \phi(x) \gg x^{\alpha}, x \to +\infty, \forall \alpha > 0. \end{cases}$$

$$(5.3)$$

Now in our present proof we have $\alpha = -\infty$ and $\alpha' = -1$, hence

$$x^{-1-\epsilon} \ll |\phi'(x)| \ll x - 1 + \epsilon, x \to +\infty, \forall \epsilon > 0,$$

and there are two a-priori contingencies about the integral $\int^{+\infty} |\phi'|$. Its divergence would imply $\phi(+\infty) = +\infty$ which cannot be; in the other case we would have

$$\int^{+\infty} |\phi'| < +\infty \Rightarrow \phi(x) = -\int_x^{+\infty} \phi'(t)\,dt = \int_x^{+\infty} |\phi'(t)|\,dt \gg \int_x^{+\infty} t^{-1-\epsilon}\,dt = \frac{x^{-\epsilon}}{\epsilon}, x \to +\infty,$$

$$(5.4)$$

which contradicts the second relation in (5.3). Notice that the procedure used to prove this last case works for any $\alpha' \in \mathbb{R}$ as well.
The last assertion in the statement of Lemma 2.3, namely "it cannot be $\alpha' > -1$", follows from the calculations in 2): $\alpha' > -1$ implies $\phi(+\infty) = +\infty$, but in this case (5.2) shows $\alpha' > -1$, a contradiction. Writting
Proof of Theorem 2.4. (I) We make explicit the assumptions writing:

$$\phi'(x)/\phi(x) = \alpha x^{-1} + o(x^{-1}), \quad \phi''(x)/\phi'(x) = (\alpha - 1)x^{-1} + o(x^{-1}), x \to +\infty, (\alpha > 0),$$

$$(5.5)$$

which in turn imply the following relations to be used in the sequel:

$$\phi(x)\phi''(x)(\phi'(x))^{-2} = \alpha^{-1}(\alpha - 1) + o(1), x \to +\infty;$$

$$(5.6)$$

$$\phi''(x)/\phi(x) = \alpha(\alpha - 1)x^{-2} + o(x^{-2}), x \to +\infty;$$

$$(5.7)$$

$$(1/\phi(x))' \sim -\alpha/x\phi(x), x \to +\infty.$$

$$(5.8)$$

First part: (2.17) Þ (2.1). If we put

$$A(x) := \left(1/\phi(x)\right)' \in AC[T, +\infty),$$

(5.9)

then, by (2.17), we may write

$$\int_T^x f = \int_T^x \frac{1}{A(t)}\left[A(t)f(t)\right]dt = -\int_T^x \frac{1}{A(t)}d\left(\int_t^{+\infty} A \cdot f\right)$$

$$= \text{constant} - \frac{1}{A(x)}\int_x^{+\infty} A(t)f(t)dt + \int_T^x \left(\frac{1}{A(t)}\right)'\left(\int_t^{+\infty} A \cdot f\right)dt.$$

(5.10)

From (5.9) and (2.17):

$$-\frac{1}{A(x)}\int_x^{+\infty} A(t)f(t)dt = \frac{\phi^2(x)}{\phi'(x)}\left[\frac{L_{f,\phi}}{\phi(x)} + o\left(\frac{1}{\phi(x)}\right)\right] = \frac{L_{f,\phi} \cdot x}{\alpha} + o(x), x \to +\infty;$$

(5.11)

$$\left(\frac{1}{A(x)}\right)'\int_x^{+\infty} A \cdot f = \left[\phi^2(x)\phi''(x)(\phi'(x))^{-2} - 2\phi(x)\right]\cdot\left[\frac{L_{f,\phi}}{\phi(x)} + o\left(\frac{1}{\phi(x)}\right)\right]$$

$$= \left[\phi(x)\phi''(x)(\phi'(x))^{-2} - 2\right]\left[L_{f,\phi} + o(1)\right] \overset{(5.6)}{=} -\alpha^{-1}(\alpha+1)L_{f,\phi} + o(1), x \to +\infty;$$

(5.12)

$$\int_T^x \left(\frac{1}{A(t)}\right)'\left(\int_t^{+\infty} A \cdot f\right)dt = -\alpha^{-1}(\alpha+1)L_{f,\phi}x + o(x), x \to +\infty.$$

(5.13)

Using (5.11) and (5.13) in the left side of (5.10) we get $\int_T^x f = -L_{f,\phi}x + o(x), x \to +\infty$, i.e. (2.1).

Second part: (2.1) Þ (2.17). First step: convergence of $\int^{+\infty} A \cdot f$. Consider the identity

$$\int_T^x A \cdot f = \int_T^x A(t)d\left(\int_T^t f\right) = \underbrace{A(x)\int_T^x f}_{I_1(x)} - \underbrace{\int_T^x A'(t)\left(\int_T^t f\right)dt}_{I_2(x)},$$

(5.14)

and estimate the behavior of $I_1(x), I_2(x)$ as $x \to +\infty$. From (2.1) and 5.8) we get:

$$A(x) = \frac{-\alpha + o(1)}{x\phi(x)}; \quad \int_T^x f = x\left[M_f + o(1)\right], x \to +\infty;$$

$$(5.15)$$

$$I_1(x) = \frac{-\alpha M_f + o(1)}{\phi(x)} = o(1), x \to +\infty.$$

$$(5.16)$$

As concerns I_2 we have:

$$\begin{cases} A(x) = -\phi'\phi^{-2}; A' = -\phi''\phi^{-2} + 2(\phi')^2 \phi^{-3}; \\ A'(x)/A(x) = (\phi''/\phi') - 2(\phi'/\phi) \overset{(5.5)}{=} -(\alpha+1)x^{-1} + o(x^{-1}), x \to +\infty, \end{cases}$$

$$(5.17)$$

from whence and (2.1) we get:

$$A'(x)\int_T^x f = A(x)\left[-(\alpha+1)M_f + o(1)\right], x \to +\infty.$$

$$(5.18)$$

As $\int_T^{+\infty} A = -\frac{1}{\phi(T)}$, we obtain the convergence of $I_2(x)$ hence, by (5.14), of $\int^{+\infty} A.f$.

Second step: asymptotic behavior of $\int_x^{+\infty} A.f$. By (5.16) and (5.18) we may integrate by parts as follows:

$$\int_x^{+\infty} A \cdot f = \int_x^{+\infty} A(t) d\left(\int_T^t f\right) = -A(x)\int_T^x f - \int_x^{+\infty} A'(t)\left(\int_T^t f\right) dt$$

$$= \frac{\alpha M_f + o(1)}{\phi(x)} + \left(\int_x^{+\infty} A\right)\left[(\alpha+1)M_f + o(1)\right]$$

$$= \frac{\alpha M_f + o(1)}{\phi(x)} - \frac{(\alpha+1)M_f + o(1)}{\phi(x)} = \frac{-M_f + o(1)}{\phi(x)},$$

$$(5.19)$$

which is (2.17) with $\int_T^x f = -L_{f,\phi} = -M_f$.

(II) From the first assumption in (2.27) we infer:

$$1/\phi \in \mathcal{R}_0(+\infty) \quad i.e. \quad A(x) = o\left(1/x\phi(x)\right), \; x \to +\infty;$$

(5.20)

and from (5.17):

$$A'(x)/A(x) = -x^{-1} + o\left(x^{-1}\right), \; x \to +\infty.$$

(5.21)

Now we retrace all steps in the second part of the proof of part (I) checking the validity of the corresponding formulas for. $\alpha = 0$ Instead of the first relation in (5.16) we have:

$$I_1(x) = o\left(1/\phi(x)\right), \; x \to +\infty,$$

(5.22)

and, instead of (5.18):

$$A'(x)\int_T^x f = A(x)\left[-M_f + o(1)\right], \; x \to +\infty.$$

(5.23)

The convergence of $\int^{+\infty} A.f$ follows as above. And using the same integration by parts as in (5.19) we get the same final relation.

I.   Let us first show that the three conditions in (2.28) imply that both $\phi, |\phi(x)|$ are rapidly-varying at $+\infty$. Conditions in $(2.28)_{1,2}$ are equivalent to $\lim_{x\to+\infty} x\phi'(x)/\phi(x) = \pm\infty$, and $(2.28)_3$ is equivalent to

$$\lim_{x\to+\infty} \phi(x)\phi''(x)/\left(\phi'(x)\right)^2 = 1,$$

(5.24)

which implies, by $(2.28)_1$, $\phi''(x) \neq 0$ ultimately; so we have:
"(5.24)" $\Leftrightarrow \phi''(x)/\phi'(x): \phi'(x)/\phi(x) \Leftrightarrow x\phi''(x)/\phi'(x): x\phi'(x)/\phi(x) \to \pm\infty, x \to +\infty.$ Now we retrace all steps in the first part of the proof of part (I) and again use decomposition (5.10); instead of (5.11) we get:

$$-\frac{1}{A(x)}\int_x^{+\infty} A \cdot f = \frac{\phi(x)}{\phi'(x)}\left[L_{f,\phi} + o(1)\right] = o(x), x \to +\infty,$$

(5.25)

and instead of (5.12) we get, using (5.24):

$$\left(\frac{1}{A(x)}\right)' \cdot \int_x^{+\infty} A \cdot f = -L_{f,\phi} + o(1), x \to +\infty,$$

(5.26)

Whence

$$\int_T^x \left(\frac{1}{A(t)}\right)' \cdot \left(\int_t^{+\infty} A \cdot f\right) dt = -L_{f,\phi}x + o(x), x \to +\infty.$$

(5.27)

From (5.25), (5.26), (5.27) we get (2.1) with $M_f = -L_{f,\phi}$
Proof of Proposition 2.6: Integration by parts gives:

$$\int_T^x t^{-\alpha}f(t)dt = \int_T^x t^{-\alpha}d\left(\int_T^t f\right) = c_1 + x^{-\alpha}\left(\int_T^x f\right) + \alpha\int_T^x t^{-\alpha-1}\left(\int_T^t f\right)dt$$

$$= \cdots \text{by } (2.1)\cdots = c_1 + M_f x^{1-\alpha} + o\left(x^{1-\alpha}\right) + \alpha M_f\left(\int_T^x t^{-\alpha}dt\right) + o\left(\int_T^x t^{-\alpha}dt\right)$$

$$= \begin{cases} c_2 + \dfrac{M_f}{1-\alpha}x^{1-\alpha} + o\left(x^{1-\alpha}\right) & \text{if } \alpha \neq 1, \\ c_3 + M_f \log x + o(\log x) & \text{if } \alpha = 1, \end{cases}$$

(5.28)

whence our claim follows dividing both sides by x.
Proof of Theorem 3: Let us assume (3.12) and start from the integral representation ([1] ; formula (6.3), p. 185):

$$f(x) = \sum_{k=1}^{n} c_k x^k + (-1)^n n! x \underbrace{\int_T^x dt_1 \cdots \int_T^{t_{n-2}} dt_{n-1} \int_T^{t_{n-1}} t^{-n-1} F_{n,0}(t) \, dt}_{n}, \quad x \geq T > 0,$$

(5.29)

which for $n = 1$ reads:

$$f(x) = cx - x \int_T^x t^{-2} F_{1,0}(t) \, dt.$$

(5.30)

From (5.30) the elementary equivalence in (3.14) easily follows, hence we suppose $n \geq 2$. If (3.12) holds true and we apply the asymptotic relation in (2.29) to $F_{n,0}$ we get:

$$\begin{cases} \int_x^{+\infty} t^{-n-1} F_{n,0}(t) \, dt = \dfrac{M_{n,0}}{n} x^{-n} + o(x^{-n}); \\[2mm] \int_T^x t^{-n-1} F_{n,0}(t) \, dt = c - \int_x^{+\infty} t^{-n-1} F_{n,0}(t) \, dt = c - \dfrac{M_{n,0}}{n} x^{-n} + o(x^{-n}); \end{cases}$$

(5.31)

and the last relation, when replaced into (5.29), yields:

$$f(x) = \sum_{k=1}^{n} c_k x^k + (-1)^n n! x \int_T^x dt_1 \, L \int_T^{t_{n-2}} \left[c - \int_{t_{n-1}}^{+\infty} t^{-n-1} F_{n,0}(t) \, dt \right] dt_{n-1}$$

$$= \sum_{k=1}^{n} \overline{c}_k x^k + (-1)^{n+1} n! x \int_T^x dt_1 \, L \int_T^{t_{n-2}} dt_{n-1} \int_{t_{n-1}}^{+\infty} t^{-n-1} F_{n,0}(t) \, dt.$$

(5.32)

But the first relation in (5.31) implies that the iterated improper integral

$\int_T^{+\infty} L \int_{t_{n-1}}^{+\infty} t^{-n-1} F_{n,0}(t) \, dt$ converges and we get a representation of type:

$$f(x) = \sum_{k=1}^{n} a_k x^k + n! x \underbrace{\int_x^{+\infty} dt_1 \cdots \int_{t_{n-1}}^{+\infty} t^{-n-1} F_{n,0}(t) \, dt}_{n}, \quad x \geq T,$$

(5.33)

together with the expansion:

$$f(x) = \sum_{k=1}^{n} a_k x^k + n!x \underbrace{\int_x^{+\infty} dt_1 \cdots \int_{t_{n-2}}^{+\infty} \left[\frac{M_{n,0}}{n} t^{-n} * o\left(t^{-n}\right) \right] dt}_{n-1} = \sum_{k=0}^{n} a_k x^k + o(1), \quad a_0 = M_{n,0},$$

(5.34)

having used one of the following elementary identities (to be used again):

$$\underbrace{\int_x^{+\infty} dt_1 \cdots \int_{t_{k-1}}^{+\infty} t^{-n} dt}_{k} = \frac{x^{k-n}}{(n-1)(n-2)\cdots(n-k)}, \quad x \geq T > 0; 1 \leq k \leq n-1.$$

(5.35)

To prove the formal differentiabilty we put:

$$I_k(x) := \begin{cases} \int_x^{+\infty} t^{-n-1} F_{n,0}(t) dt & \text{for } k = 0; \\ \int_x^{+\infty} I_{k-1}(t) dt & \text{for } 1 \leq k \leq n-1; \end{cases}$$

(5.36)

and from (5.31) we infer relations:

$$I_k(x) = \begin{cases} \dfrac{M_{n,0}}{n} x^{-n} + o\left(x^{-n}\right) & \text{for } k = 0; \\ \dfrac{M_{n,0} x^{-n+k}}{(n-1)(n-2)L\ (n-k)} + o\left(x^{-n+k}\right) & \text{for } 1 \leq k \leq n-1. \end{cases}$$

(5.37)

Calling $P_n(x)$ the last sum on the right in (5.34), which differ by a constant from the sum on the right in (5.33), and applying Leibniz's rule to (5.33) we get:

$$f^{(k)}(x) = P_n^{(k)}(x) + (-1)^k n! x I_{n-k-1}(x) + (-1)^{k-1} n! k I_{n-k}(x)$$

$$= P_n^{(k)}(x) + \frac{(-1)^k n! M_{n,0} x^{-k}}{(n-1)(n-2)L\ (k+1)} + \frac{(-1)^{k-1} n! k M_{n,0} x^{-k}}{(n-1)(n-2)L\ (k+1)k} + o\left(x^{-k}\right)$$

$$= P_n^{(k)}(x) + o\left(x^{-k}\right) \text{ for } 1 \leq k \leq n-1.$$

(5.38)

The expressions of $f^{(n-1)}$ and $f^{(n)}$ involve $I_0(x)$ and its derivative:

$$\begin{cases} f^{(n-1)}(x) = P_n^{(n-1)}(x) + (-1)^{n-1} n! x I_0(x) + (-1)^n n!(n-1) I_1(x); \\ f^{(n)}(x) = n! a_n + (-1)^{n-1} n! I_0(x) + (-1)^n n! x^{-n} F_{n,0}(x) + (-1)^{n-1} n!(n-1) I_0(x) \\ \qquad = n! a_n + (-1)^{n-1} n! n I_0(x) + (-1)^n n! x^{-n} F_{n,0}(x). \end{cases}$$

$$(5.39)$$

So far we have proved that (3.12) implies relations in (3.13) for $f^{(k)}, 1 \leq k \leq n-1$, without any information on $f^{(n)}$, and, for the time being, P_n is a non-better specified polynomial of degree $\leq n$. To prove (3.11) we estimate the behavior, as $x \to \pm\infty$, of $f_{n,k}(x)$ for $1 \leq k \leq n$ using its known expression in terms of f, ([1] ; formula (2.6), p. 168):

$$F_{n,k}(x) = \frac{1}{k!} \sum_{i=0}^{n-k} \frac{(-x)^i}{i!} f^{(k+i)}(x) = L \text{ by } (5.38),(5.39) L$$

$$= \frac{1}{k!} \sum_{i=0}^{n-k} \frac{(-x)^i}{i!} P_n^{(k+i)}(x) + \frac{1}{k!} \sum_{i=0}^{n-k-1} \frac{(-x)^i}{i!} f^{(k+i)}(x) + \frac{(-x)^{n-k}}{k!(n-k)!} \left[f^{(n)}(x) - P_n^{(n)}(x) \right] = L$$

$$(5.40)$$

as the first sum is nothing but the expression of the coefficient of the power x^k in the polynomial P_n, i.e. a^k.

$$L = a_k + \frac{1}{k!} \sum_{i=0}^{n-k-1} \left[\frac{(-1)^i n!}{i!} x^{i+1} I_{n-k-i-1}(x) + \frac{(-1)^{k-1} n! k}{i!} x^i I_{n-k-i}(x) \right]$$

$$+ \frac{(-1)^{k-1} n! n}{k!(n-k)!} x^{n-k} I_0(x) + \frac{(-1)^k n!}{k!(n-k)!} x^{-k} F_{n,0}(x) = L \text{ by } (5.37) L$$

$$= a_k + \sum_{j=0}^{n-k-1} O(x^{-i}) + O(x^{-k}) + \frac{(-1)^k n!}{k!(n-k)!} x^{-k} F_{n,0}(x)$$

$$= a_k + \frac{(-1)^k n!}{k!(n-k)!} x^{-k} F_{n,0}(x) + o(1).$$

By (2.34) the function $x^{-k} F_{n,0}(x)$ has asymptotic mean "zero" and the same is true for a term $o(1)$; so the sum of the last three terms above represents a function with asymptotic mean equalling a_k. We have proved that "2) Þ 1) Ù 3)". It remains to show "3) Þ 2)". First step. Let us first evaluate $f^{(n)}$ from representation (5.29); putting

$$J_k(x) := \begin{cases} \int_T^x t^{-n-1} F_{n,0}(t)\, dt & \text{for } k = 0; \\ \int_T^x J_{k-1}(t)\, dt & \text{for } 1 \le k \le n-1; \end{cases}$$

(5.41)

we get:

$$f^{(n)}(x) = n!c_n + (-1)^n n! \left[n J_0(x) + x^{-n} F_{n,0}(x) \right].$$

(5.42)

Now we start as in (5.40) from the expression of $F_{n,0}$:

$$F_{n,0}(x) = \sum_{k=0}^{n} \frac{(-x)^k}{k!} f^{(k)}(x) = L \text{ by } (3.13) L$$

$$= \sum_{k=0}^{n-1} \frac{(-x)^k}{k!} \left[P_n^{(k)}(x) + o\left(x^{-k}\right) \right] + \frac{(-x)^n}{n!} f^{(n)}(x) = L \text{ by } (5.42) L$$

$$= \sum_{k=0}^{n-1} \frac{(-x)^k}{k!} P_n^{(k)}(x) + o(1) + (-x)^n c_n + n x^n J_0(x) + F_{n,0}(x),$$

(5.43)

whence we get

$$J_0(x) = \sum_{k=0}^{n-1} \frac{(-1)^k}{k!\,n} x^{k-n} P_n^{(k)}(x) + \frac{c_n}{n} + o\left(x^{-n}\right) = \frac{c_n}{n} + o(1),$$

(5.44)

which implies the convergence of the improper integral $\int_T^{+\infty} t^{-n-1} F_{n,0}(t)\, dt$; and we can rewrite representation (5.29) in the form:

$$f(x) = \sum_{k=1}^{n} \bar{c}_k x^k + (-1)^{n+1} n! x \int_T^x dt_1 \, L \int_T^{t_{n-2}} dt_{n-1} \int_{t_{n-1}}^{+\infty} t^{-n-1} F_{n,0}(t)\, dt, \quad x \ge T > 0,$$

$$= \sum_{k=1}^{n} \bar{c}_k x^k + o(x^n).$$

(5.45)

Comparing (5.45) and the assumed relation $f(x) = p_n(x) + o(1)$ we infer that the two polynomials p_n and the sum appearing in (5.45) have the same leading coefficient: $a_n = \overline{c}_n$. Now we do calculations just like those from (5.41) to (5.43) but starting from representation (5.45) and paying attention to the signs, so getting:

$$f^{(n)}(x) = n!a_n + (-1)^{n+1} n!n\int_x^{+\infty} t^{-n-1}F_{n,0}(t)\,dt + (-1)^n n!x^{-n}F_{n,0}(x);$$

(5.46)

$$F_{n,0}(x) = \sum_{k=0}^n \frac{(-x)^k}{k!}P_n^{(k)}(x) + (-1)^{n+1} n!n\int_x^{+\infty} t^{-n-1}F_{n,0}(t)\,dt + (-1)^n n!x^{-n}F_{n,0}(x)$$

$$= a_0 - nx^n\int_x^{+\infty} t^{-n-1}F_{n,0}(t)\,dt + F_{n,0}(x) + o(1),$$

(5.47)

having used the identity $\sum_{k=0}^n \frac{(-x)^k}{k!}P_n^{(k)}(x) \equiv \alpha_0$, ([1] ; Lemma 2.2, p. 169). From (5.47) we infer

$$\int_x^{+\infty} t^{-n-1}F_{n,0}(t)\,dt = \frac{a_0}{n}x^{-n} + o(x^{-n}),$$

(5.48)

which, by (2.29), implies $F_{n,0} \in M_\infty$ and $\alpha_0 = M_{n,0}$.

Proof of Theorem 3.4. The only thing to be proved is that an expansion (3.10) plus condition (3.16) imply $F_{n,0} \in M_\infty$. We first show that it is enough to prove our claim with (3.16) replaced by the condition of one-signedness:

either $F_{n,0}(x) \leq 0 \ \forall x \geq T$ or $0 \leq F_{n,0}(x) \ \forall x \geq T.$

(5.49)

In fact it is known, ([1] ; Lemma 2.2, p. 169), that: $F_{n,0} = c = \text{cons}\tan t$ iff f is a polynomial of type

$$p(x) = a_n x^n + L + a_1 x + c.$$
(5.50)

Let now g be any function, $g \in AC^{n-1}[T,+\infty)$, let p be a polynomial of type (5.50) and define: $f(x) := g(x) - p(x)$. With an obvious meaning of the symbol $G_{n,0}$ we have: $F_{n,0} = G_{n,0} - c$; hence:

$$F_{n,0} Z 0 \Leftrightarrow G_{n,0} \quad c.$$
(5.51)

It follows that any result on formal differentiability of a polynomial asymptotic expansion involving g admits of a literal transposition to a polynomial asymptotic expansion involving f. Our assumption are now: expansion (3.10) and one-signedness of $F_{n,0}$, and the proof (which we make explicit here) is a word-for-word repetition of that in ([1] ; Proof of Th. 4.2, pp. 193-195) with a slight modification at the conclusive passage. From representation (5.29) we infer

$$x^{-n} f(x) = c_n + o(1) + (-1)^n \, n! \, x^{1-n} \int_T^x dt_1 \, L \int_T^{t_{n-2}} dt_{n-1} \int_T^{t_{n-1}} t^{-n-1} F_{n,0}(t) \, dt,$$
(5.52)

and, by (3.10), the following limit:

$$\lim_{x \to +\infty} \left(\int_T^x dt_1 \, L \int_T^{t_{n-2}} dt_{n-1} \int_T^{t_{n-1}} t^{-n-1} F_{n,0}(t) \, dt \right) \Big/ x^{n-1} \quad \text{exists as a finite number.}$$
(5.53)

For $n = 1$ (3.10) reduces to $f(x) = a_1 x + a_0 + o(1)$ and (5.53) is " $\int_T^{+\infty} t^{-2} F_{1,0} dt$ convergent". Hence representation (5.29) can be rewritten in the form $f(x) = a_1 x + \overline{a}_0 + x \int_T^{+\infty} t^{-2} F_{1,0} dt$, and (3.10) implies that " $\lim_{x \to +\infty} x \int_T^{+\infty} t^{-2} F_{1,0} dt$ exists in $\mathbb{R}$ " which is equivalent to $F_{n,0} \in M_\infty$.

For $n \geq 2$ we apply L'Hospital's rule $(n-1)$ times to the limit in (5.53) so getting the limit:

$$\lim_{x \to +\infty} \int_T^x t^{-n-1} F_{n,0}(t) \, dt \big/ (n-1)!.$$

(5.54)

By the one-signedness of $F_{n,0}$ this last limit exists in the extended real line, hence it must be a finite number. This means the convergence of $\int_T^{+\infty} t^{-n-1} F_{n,0}(t) dt$ and representation (5.29) can be rewritten as:

$$f(x) = \sum_{k=1}^{n} c_k x^k + (-1)^{n+1} n! x \int_T^x dt_1 \, L \int_T^{t_{n-2}} dt_{n-1} \int_{t_{n-1}}^{+\infty} t^{-n-1} F_{n,0}(t) \, dt = \sum_{k=1}^{n} c_k x^k + o(x^n).$$

(5.55)

The last relation implies that c^n coincides with the a^n in (3.10) and we get:

$$x^{1-n} \left(f(x) - a_n x^n \right) = c_{n-1} + o(1) + (-1)^{n+1} n! x^{2-n} \int_T^x dt_1 \, L \int_T^{t_{n-2}} dt_{n-1} \int_{t_{n-1}}^{+\infty} t^{-n-1} F_{n,0}(t) \, dt.$$

(5.56)

By the above argument involving L'Hospital's rule we arrive at the convergence of the iterated integral

$$\int_T^{+\infty} d\tau \int_\tau^{+\infty} t^{-n-1} F_{n,0}(t) dt.$$ An iteration of the procedure yields condition

$$\int_T^{+\infty} dt_1 \, L \int_{t_{n-2}}^{+\infty} dt_{n-1} \int_{t_{n-1}}^{+\infty} t^{-n-1} F_{n,0}(t) \, dt \quad \text{convergent,}$$

(5.57)

which implies representation

$$f(x) = a_n x^n + L + a_1 x + n! x \int_x^{+\infty} dt_1 \, L \int_{t_{n-2}}^{+\infty} dt_{n-1} \int_{t_{n-1}}^{+\infty} t^{-n-1} F_{n,0}(t) \, dt,$$

(5.58)

where the coefficients a_k are those in (3.10). From (5.58) we infer that

$$\lim_{x \to +\infty} \left(\int_x^{+\infty} dt_1 \, L \int_{t_{n-2}}^{+\infty} dt_{n-1} \int_{t_{n-1}}^{+\infty} t^{-n-1} F_{n,0}(t)\,dt \right) \Big/ x^{-1} = a_0,$$

$$(5.59)$$

and applications of L'Hospital's rule $(n-1)$ times yields the limit

$$\lim_{x \to +\infty} \left(\int_x^{+\infty} t^{-n-1} F_{n,0}(t)\,dt \right) \Big/ (n-1)! \, x^{-n} = a_0,$$

$$(5.60)$$

which, by (2.29), is equivalent to $F_{n,0} \in M_\infty$.
In passing notice that the last calculations and (5.34) prove that:

For a given function $g \in L^1_{loc}[T, +\infty)$ and g one-signed the following equivalence holds true:

$$\{ g \in M_\infty \text{ and } M_g = 1 \} \Leftrightarrow \int_x^{+\infty} dt_1 \, L \int_{t_{n-2}}^{+\infty} dt_{n-1} \int_{t_{n-1}}^{+\infty} t^{-n-1} g(t)\,dt = \frac{1}{n!\,x} + o\left(\frac{1}{x}\right), \; x \to +\infty.$$

$$(5.61)$$

REFERENCES

1. Granata, A. (2007) Polynomial Asymptotic Expansions in the Real Domain: The Geometric, the Factorizational, and the Stabilization Approaches. Analysis Mathematica, 33, 161-198. http://dx.doi.org/10.1007/s10476-007-0301-0.
2. Granata, A. (2010) The Problem of Differentiating an Asymptotic Expansion in Real Powers. Part I: Unsatisfactory or Partial Results by Classical Approaches. Analysis Mathematica, 36, 85-112. http://dx.doi.org/10.1007/s10476-010-0201-6.
3. Granata, A. (2010) The Problem of Differentiating an Asymptotic Expansion in Real Powers. Part II: Factorizational Theory. Analysis Mathematica, 36, 173-218. http://dx.doi.org/10.1007/s10476-010-0301-3.
4. Granata, A. (2011) Analytic Theory of Finite Asymptotic Expansions in the Real Domain. Part I: Two-Term Expansions of Differentiable Functions. Analysis Mathematica, 37, 245-287. (For an Enlarged Version with Corrected Misprints see: arxiv.org/abs/1405.6745v1 mathCA.. http://dx.doi.org/10.1007/s10476-011-0402-7.
5. Granata, A. (2014) Analytic Theory of Finite Asymptotic Expansions in the Real Domain. Part II: The Factorizational Theory for Chebyshev Asymptotic Scales. Electronically Archived—arXiv: 1406.4321v2 math.CA.

6. Granata, A. (2015) The Factorizational Theory of Finite Asymptotic Expansions in the Real Domain: A Survey of the Main Results. Advances in Pure Mathematics, 5, 1-20. http://dx.doi.org/10.4236/apm.2015.51001.
7. Haupt, O. (1922) Über Asymptoten ebener Kurven. Journal für die Reine und Angewandte Mathematik, 152, 6-10; ibidem, 239.
8. Sanders, J.A. and Verhulst, F. (1985) Averaging Methods in Nonlinear Dynamical Systems. Springer-Verlag, New York.
9. Corduneanu, C. (1968) Almost Periodic Functions. Interscience Publishers, New York.
10. Faedo, S. (1946) Il Teorema di Fuchs per le Equazioni Differenziali Lineari a Coefficienti non Analitici e Proprietà Asintotiche delle Soluzioni. Annali di Matematica Pura ed Applicata (the 4th Series), 25, 111-133. http://dx.doi.org/10.1007/BF02418080.
11. Hallam, T.G. (1967) Asymptotic Behavior of the Solutions of a Nonhomogeneous Singular Equation. Journal of Differential Equations, 3, 135-152. http://dx.doi.org/10.1016/0022-0396(67)90011-3.
12. Hukuhara, M. (1934) Sur les Points Singuliers des Équations Différentielles Linéaires; Domaine Réel. Journal of the Faculty of Science, Hokkaido University, Ser. I, 2, 13-88.
13. Ostrowski, A.M. (1951) Note on an Infinite Integral. Duke Mathematical Journal, 18, 355-359. http://dx.doi.org/10.1215/S0012-7094-51-01826-1.
14. Agnew, R.P. (1942) Limits of Integrals. Duke Mathematical Journal, 9, 10-19. http://dx.doi.org/10.1215/S0012-7094-42-00902-5.
15. Hardy, G.H. (1911) Fourier's Double Integral and the Theory of Divergent Integrals. Transactions of the Cambridge Philosophical Society, 21, 427-451.
16. Hardy, G.H. (1949) Divergent Series. Oxford University Press, Oxford. (Reprinted in 1973).
17. Blinov, I.N. (1983) Absence of Exact Mean Values for Certain Bounded Functions. Izvestija Akademii Nauk SSSR. Serija Mathematicheskaja (Moscow), 47, 1162-1181.
18. Ditkine, V. and Proudnikov, A. (1979) Calcul Opérationnel. Éditions Mir, Moscou.
19. Baumgärtel, H. and Wollenberg, M. (1983) Mathematical Scattering Theory. Birkhäuser Verlag, Berlin.
20. Ostrowski, A.M. (1976) On Cauchy-Frullani Integrals. Commentarii Mathematici Helvetici, 51, 57-91. http://dx.doi.org/10.1007/BF02568143.
21. Bingham, N.H., Goldie, C.M. and Teugels, J.L. (1987) Regular Variation. Cambridge University Press, Cambridge. http://dx.doi.org/10.1017/CBO9780511721434
22. Hartman, Ph. (1952) On Non-Oscillatory Linear Differential Equations of Second Order. American Journal of Mathematics, 74, 389-400. http://dx.doi.org/10.2307/2372004.
23. Hartman, Ph. (1982) Ordinary Differential Equations. 2nd Edition, Birkhäuser, Boston.

24. Giblin, P.J. (1972) What Is an Asymptote? The Mathematical Gazette, 56, 274-284. http://dx.doi.org/10.2307/3617830.

Citation

Granata, A. (2015) The Role of Asymptotic Mean in the Geometric Theory of Asymptotic Expansions in the Real Domain. Advances in Pure Mathematics, 5, 100-119. doi: 10.4236/apm.2015.52013.

Index